PERGAMON SERIES OF MONOGRAPHS IN
FURNITURE AND TIMBER
GENERAL EDITOR: JACK CAPE, A.R.C.A., A.T.D., F.S.A.E.

VOLUME 12

THE TIMBER TRADE—AN INTRODUCTION TO COMMERCIAL ASPECTS

PERGAMON SERIES OF MONOGRAPHS IN
FURNITURE AND TIMBER

THE TIMBER TRADE –
An Introduction
to Commercial Aspects

JACK H. LEIGH, A. I. W. Sc.

PERGAMON PRESS

Oxford · New York

Toronto · Sydney · Braunschweig

Pergamon Press Ltd., Headington Hill Hall, Oxford
Pergamon Press Inc., Maxwell House, Fairview Park, Elmsford,
New York 10523
Pergamon of Canada Ltd., 207 Queen's Quay West, Toronto 1
Pergamon Press (Aust.) Pty. Ltd., 19a Boundary Street,
Rushcutters Bay, N.S.W. 2011, Australia
Vieweg & Sohn GmbH, Burgplatz 1, Braunschweig

First edition 1971

Library of Congress Catalog Card No. 70-133797

Printed in Germany
08 016335 1

1650919

CONTENTS

LIST OF TABLES

PREFACE

THIS book is written as a guide to the commercial aspects of the Timber Trade. It is not intended that it should attempt to cover the complexities of forestry, wood technology and utilization. So many excellent textbooks are available on such subjects and one could never hope to do justice to so wide a field in one small volume.

The information contained in the following chapters refers, in the main, to the United Kingdom trade. However, since this consists almost exclusively of imports from other timber-producing countries, it follows that the general structure and procedure of the trade as given are universally accepted.

J. H. L.

GEOGRAPHICAL OUTLINE

1. THE WORLD'S RESOURCES

The Timber Trade may well be regarded as devised to bring the forest resources of the world to those who wish to have them and in the form in which they would be most useful.

Wood is universally abundant and there is scarcely any country in the world in which trees do not grow, but some countries do not have sufficient for their own needs whilst others have more than they need. Additionally, the kind of tree varies very much, depending on geographical position and various other factors, which means that an interchange of supplies becomes necessary. This represents a state of affairs which requires a complex organization for handling to the best advantage.

Forests cover approximately 25 per cent of the world's land area—8000 million acres. These forests could be described as inexhaustible because, unlike the resources of mineral raw materials which become depleted, the forest is not a mine but a cropland. Provided that it is harvested as crops should be, with skilled management and long-term planning, then its yield is sufficient to satisfy the requirements of the world indefinitely. Unfortunately, in past years this forestry control has not been too skilfully planned, with the result that many areas became wasteland or desert. This is not true today, when timber-producing countries appreciate their source of wealth and have adequate means of ensuring its continuance. Help from organizations such as F.A.O. is available for all and teams of experts are continually advising the more backward countries who do not possess efficient forestry and utilization facilities.

It is estimated that about 3000 million acres of forest are being commercially exploited, which leaves vast areas relatively untouched. Economic considerations involving problems of extraction, transportation, communications and shipment are contributory factors to this untapped source. Nevertheless, these considerations are being overcome by the ever-increasing demands for wood products and this is illustrated by the figures shown in Table I.

TABLE I. REMOVALS OF WOOD FROM FORESTS

(Million cubic metres)

Region	Total			Fuel Wood			Industrial Wood		
	1950	1960	1968	1950	1960	1968	1950	1960	1968
Europe	287	316	308	114	104	68	173	212	240
U.S.S.R.	278	373	380	112	111	91	166	262	290
N. America	367	399	499	70	51	27	297	348	471
C. America	33	43	44	28	36	34	5	7	10
S. America	150	171	229	132	144	195	18	27	35
Africa	96	181	247	89	163	214	7	18	33
Asia	215	304	406	165	173	261	50	131	144
Pacific Area	17	24	26	7	9	7	10	15	19
World total	1443	1811	2139	717	790	897	726	1021	1242

These figures show that the world total removals increased with the exception of Europe. The growing productivity of the world's forests was common in most regions and in the years 1950 to 1968, the greatest increase was in Asia, North America and Africa, in that order.

Some interesting facts are apparent. In 1950, half of the wood was used as fuel and this form of utilization steadily declined in all areas in the northern hemisphere. In terms of consumption *per capita*, this is associated with the advance of industrial development which has introduced higher standards of living with alternative means of providing energy. Associated with this is the increase of industrial wood removals which is accounted for by the greater demand not only for constructional wood but for wood pulp and allied wood products.

2. FOREST AREAS

General

Over 40,000 species of woody plants have been botanically identified and of these some 10,000 are trees of some economic importance—at least in their local environment. Obviously, only a minority of these are of commercial importance in terms of world trade and there are many factors which influence quality of growth and resulting value. Among these are the variations between a continental and an oceanic climate, the effect of altitude which tends to compensate for latitude, rainfall, temperature and the effects of mountain ranges, deserts and the Gulf Stream. Moreover, many tree genera are represented by numerous species, which between them have a very wide natural distribution. For example, there are some eighty species of pine occurring within a very big range of latitude, and the Scots Pine (*Pinus sylvestris*), a native of Scotland, grows in the "conifer belt" across the north-eastern hemisphere and is known by many commercial trade names, e.g. Archangel Redwood, Swedish Redwood, etc.

The forest areas of the world consist, in broad terms, of

(a) Softwoods—conifers.
(b) Softwoods and hardwoods—mixed.
(c) Hardwoods—temperate.
(d) Hardwoods—tropical.

Figure 1 illustrates these areas. Before describing these areas in greater detail, it may be advisable to define more

3

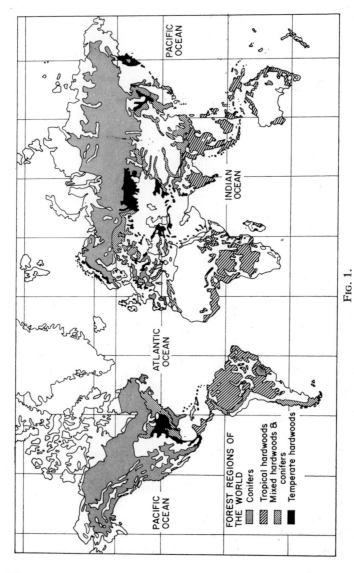

FIG. 1.

FOREST REGIONS OF
THE WORLD
░ Conifers
▨ Tropical hardwoods
▧ Mixed hardwoods &
conifers
■ Temperate hardwoods

clearly what is meant by softwood and hardwood. The designation is a poor one and has no connection with the density of the wood. It is a broad classification which is commercially understood, but it gives rise to many anomalies as some softwoods are hard, such as yew, and some hardwoods are soft, such as balsa.

Softwoods are derived from a class of trees known as conifers (Coniferae), which are cone-bearing and have needle or scale-like leaves and are usually evergreen.

Hardwoods, on the other hand, come from a class of trees which bear broad leaves, i.e. a definite flat surface area. These trees are more highly developed plants than the conifers, and whereas species of the latter are numbered in hundreds, there are many thousand species of broadleaved trees. In temperate climates, such trees are usually deciduous, but there are evergreen hardwoods.

Softwoods

The evergreen coniferous forest belt stretches across the northern area of the world through Scotland, Scandinavia, the Baltic countries, Russia and Siberia. On the other side of the Atlantic, from Eastern Canada to the Rockies and down the Pacific Coast from Alaska to California. The upper edge of this belt marks the northern limit of tree growth, "the timber line", where, in the extreme north, trees become no more than prostrate shrubs and give way to mosses and lichens and finally to the eternal snows. It is from this conifer belt that the bulk of the world's softwood is obtained.

Secondly, the southern edge of the coniferous forest merges gradually and over a wide area into the mixed softwood and hardwood forests of the north temperate zone. This is typical of the forests in the British Isles, France, Germany and Central Europe.

Thirdly, in the subtropical or "Mediterranean" climatic zone, a predominantly evergreen vegetation is found and this natural range extends even further south in mountainous regions where the climate is comparable to that found in northern latitudes. There are the cedars of the Atlas Mountains, pines and yews of the Himalayas and conifers at high altitudes as far south as Mexico.

The Conifer Belt—Eastern Hemisphere

In this area, two species predominate, the European Redwood (*Pinus sylvestris*) and European Whitewood (*Picea abies*).

European Redwood is the standard name in the British Isles when imported, the home-grown wood being called Scots Pine. Other names include Baltic, Finnish, Swedish, Archangel, Siberian and Polish Redwood and Yellow or Red Deal. (It should be noted that deal refers to a size and this description is therefore a misnomer.) This wood is the most important structural wood in Europe and the principal exporting countries are Russia, Finland and Sweden, in that order.

European Whitewood is the standard name in the British Isles when imported, the home-grown wood being called European Spruce or Common Spruce. Other names include Baltic, Finnish, Swedish, Russian, Yugoslavian and Czecho-slovakian Whitewood, Norway Spruce and White Deal. (Not to be confused with American Whitewood, *Liriodendron*, which is a hardwood.) This wood is used for much the same purposes as redwood and, in addition, is important as a raw material for the manufacture of paper pulp. It is universally familiar as the traditional Christmas tree.

In company with these two trees are found the larches, aspen and European birch. The latter is one of the few hard-wood trees in addition to aspen which can thrive in the austere conditions natural to the conifers.

The Conifer Belt—Western Hemisphere

In Eastern Canada, the forests are very similar in appearance to those previously described. Pine, spruce and birch again predominate although of different species from those of the eastern hemisphere. Among the pines, Red Pine (*Pinus resinosa*) is perhaps the most important whereas, at one time, Weymouth or Yellow Pine (*Pinus strobus*) held pride of place. Forest fires, insect damage and heavy felling have reduced the supplies of this very fine wood, which is prized for pattern making and joinery. Jack Pine, Balsam Fir and Eastern Hemlock, together with three or four species of spruce, constitute the rest of the forests and the latter are so similar that they are generally marketed together as Canadian, Quebec, New Brunswick and Nova Scotia Spruce. This most important spruce is of interest in that it is found growing further north than almost any tree and extends in range right across the Dominion to the Rocky Mountains and up into Alaska.

Finally, in the Rockies and on the Pacific Coast, from Alaska to California, are found some of the finest and most valuable forests in the world, certainly the finest coniferous forests. The trees themselves for the most part are longer lived and attain much greater dimensions than those in any other part of the conifer belt. There are Douglas Firs and Western Red Cedar, trees of virgin growth and many of them centuries old and from 200 to 300 ft tall.

First in importance is the Douglas Fir (*Pseudotsuga taxifolia*), which is the largest, tallest and most abundant tree in all this wealth of coniferous forest. It is still known also as Oregon and British Columbian Pine, but Douglas Fir is the standard name. The tree itself, which is, in fact, neither a true fir nor a true pine, belongs to a separate and distinct genus, *Pseudotsuga*, and produces one of the world's most important and widely used structural timbers employed for many of the same purposes as European Redwood, although, of course, it is obtainable in very much larger dimensions. The other woods of commercial importance are Western Hemlock (*Tsuga heterophylla*), Western Red Cedar (*Thuya plicata*), Sitka Spruce (*Picea sitchensis*), Western White Pine (*Pinus monticola*), a variety of spruces and white firs, Port Orford Cedar (*Cupressus lawsoniana*) and the Balsam, which is frequently mixed with

7

Hemlock. Over the border there are the Western pines—Idaho Pine, which is identical with the *Pinus monticola* of British Columbia, Ponderosa Pine (*Pinus ponderosa*) and Sugar Pine (*Pinus lambertiana*)—as an export wood, though now rare, the Californian Redwood or Sequoia (*Sequoia sempervirens*). Although not of any commercial importance, it is interesting to note that at one time, before the last ice age, this tree was to be found in Europe and in parts of Asia as traces of them have been found in the coal measures. When the ice receded the ancestors of the existing trees were all that had survived and these trees are probably today the oldest and largest living things in the world. Many of them are above 300 ft in height and 30 ft in diameter and their age is estimated as being between 3000 and 4000 years, possibly more in some cases.

At one time, the southern states of the United States provided a wealth of fine-quality Pitch Pine, but partly because the fashion has changed, and partly because of the decreasing amount of really high-grade wood, not so very much is seen today. There are three types of Pitch Pine, or, as it is known in North America, Southern Pine or Yellow Pine, and they are Long Leaf (*Pinus palustris*), Short Leaf (*Pinus echinata*) and Loblolly (*Pinus taeda*).

Hardwoods

As the evergreen coniferous forests extend south they tend to merge with the temperate hardwoods in more southerly latitudes. There is, of course, no east-to-west line at which the conifers stop short and where their place is taken by the broad-leaved trees. For example, pine, spruce, oak, ash and maple are found growing together over a wide area and the latter three hardwoods only begin to preponderate gradually as the southern fringes of the conifer's natural range are approached.

8

The Temperate Hardwood Forests

The woodlands of the British Isles are typical of the temperate hardwood forests. There is greater variety of species than in the coniferous forests, although trees such as beech do sometimes form more or less pure woods in some areas. In the Chilterns, for example, the soil is particularly suitable for beech, where the trees have gradually attained complete ascendancy by virtue of the dense shade cast by their foliage, in which few other plants can grow, and by its own ability to thrive in the shade of other species during the long struggle for supremacy.

In the forests of the British Isles and Europe are found the oak, ash, elm, beech, sweet chestnut, walnut and sycamore, which represent the most important commercial species. Numerous others, such as alder, willow, lime, cherry and poplar, are of use for particular purposes.

In the Eastern United States and Southern Ontario, other species of these trees are found, with the addition of maple, hickory, red gum and the tulip tree which produces the timber known as Canary White Wood in the British Isles and as Yellow Poplar in the United States.

There are hundreds of species of trees to be found throughout the north temperate regions but few of these are of real commercial significance.

The Tropical Forests

These forests are the most extensive and varied of all and it is no exaggeration to say that they involve vast areas which are unexplored and doubtless contain species of trees which are virtually unknown. Broadly speaking, the timbers from tropical trees can be divided into three main groups:

(1) Soft and light in weight, e.g. Obeche from West Africa, about 24 lb/cu. ft.
(2) Medium hard and heavy, e.g. Teak from Burma and Siam, about 43 lb/cu. ft.
(3) Very hard and heavy, e.g. Greenheart from British Guiana, about 65 lb/cu. ft.

9

There are enormous numbers of timbers of each type and every year brings new species into commercial use.

Perhaps the most important, certainly the best known, are the Mahoganies, the large majority of which fall within the medium hard and heavy category. The Mahogany family is composed of many genera, each of a number of species, and is represented in most parts of the tropics. The original Spanish Mahogany from Cuba and other West Indian islands is now rarely exported, supplies having been almost exhausted by the demands of the last two centuries. Its place has been taken by other species, Central American or Honduras Mahogany, and by various African Mahoganies which are of different genera. Many woods, excellent in their own right but not true Mahoganies, have acquired the name Mahogany owing to some superficial resemblance.

It is quite impossible to list the commercially important timbers from the various hardwood regions of the world. From West Africa alone, some seventy-five different species are exported and, as previously mentioned, the list grows every year as certain areas become temporarily depleted and fresh areas are worked. Briefly, the following timbers are of commercial importance from the various exporting areas of the world.

In West Africa, the more coastal areas consist of dense tropical forests whereas inland these tend to become of the open forest type with accompanying grasslands. Abura, agba, avodire, ekki, guarea, iroko, mahogany, makore, mansonia, obeche, sapele and utile are among those which are universally utilized.

In South America the forest types are similar to West Africa but there are greater areas of inaccessibility. Balsa, boxwood, cedar, greenheart, laurel, lignum vitae, mahogany, peroba, purpleheart, rosewood and zebrawood are the most important timbers. (Parana Pine, a softwood, is the best-known export of all.)

South-east Asia produces a considerable number of extremely important commercial timbers. The area is vast and the forest types range from tropical rain and mangrove swamps to savannah and evergreen rain forests. Dhup, ebony, eng, gurjun/keruing/yang, jelutong, kapur, lauan, laurel, meranti, oak, padauk, pyinkado, ramin, rosewood, seraya, and teak, these are only a few of the vast number of valuable species which are available in considerable quantities from this area.

In the southern hemisphere, Australia and New Zealand provide a wealth of timber, the former country being particularly noted for the eucalyptus. This genus comprises about 500 species, the best known being karri, jarrah, tallowwood, blue gum and Tasmanian oak. Other than the Eucalypts are black bean, walnut and silky oak.

THE STRUCTURE OF THE TIMBER TRADE

1. GENERAL

As a matter of essential convenience, the trade is divided fundamentally into softwoods, hardwoods, plywood, fibre boards, homegrown, etc., although many commercial concerns may produce or trade in several or all of them. Apart from the purely commercial aspect of selling and buying, these groups are represented by trade organizations whose functions are to protect, promote and develop common interests, to support research and development and to provide publicity.

There are many hundreds of species commercially involved in world trade. When it is considered that timber is not a homogenous product but varies in quality from piece to piece together with a very wide range of sizes, it is not surprising that a complex organization is required to effect transfer from the forest to the ultimate consumer who may well be on the other side of the world. Each country has its own custom of the trade, but in general terms the pattern is the same, with only minor changes to suit specific local requirements. The structure given in the following pages has existed for at least a century and that in itself is some proof of its efficiency and satisfaction for all the needs of the various sections of the trade.

The three main classes of trader are the overseas principal, known as the shipper, the shipper's agent, who is normally a resident in the importing country, and the importer, who is normally the first owner of the goods. The importer may sell direct to the consumer or to a fourth class of trader, the non-importing merchant, who in turn sells to the consumer.

2. THE SHIPPER

The shipper is the sawmill owner and ships the timber, and should not be confused with the ship owner. Usually he buys his logs from the producers or forest owners, although many shippers do own their own forests and this is particularly so in the hardwood trade. The logs are conveyed to his sawmill where they are sawn to the exact measurements required by the ultimate user in the receiving countries. In certain instances, such as the African hardwood trade, the logs are often shipped without conversion.

The shipper aims to dispose of his complete output to the best advantage and to avoid having the poorer qualities and less popular dimensions left on his hands. Most shippers have their accredited agents in the countries to which they export and some appoint sole agents and others employ several. The shipper passes on to his agent full particulars about the timber he has to offer at any one time, describing the goods in terms both of dimension and quality, using the various recognized classifications of grading (see Chapter 4, Grading of Timber). He requires his agent to keep him informed about market possibilities, to arrange his sales and to guarantee him against risk of loss through a buyer's inability to meet his financial commitments.

Shippers have their own associations which look after their common interests and which negotiate any procedure of an international nature such as contract form conditions, etc. Depending upon the type of contract negotiated, shippers may also have to arrange for the chartering of a vessel to carry the goods to the country of destination and to arrange the insurance (see Chapter 3, section 2, Contracts).

3. THE AGENT

The timber trade agent performs many functions. He is undoubtedly the professional man in the matter of distribution and is free to devote his entire time to the intelligent marketing of the shipper's production, or that part of it which is suitable for his country's needs, at the same time protecting the interests of both sellers and buyers.

First, he has to place the stock offered by his shipper, to whom he owes primary responsibility. As far as possible, he has to find buyers who, individually or collectively, are prepared to take a reasonably full range of the sizes and qualities which his principal has to offer. Occasions often arise where special specifications are required by a buyer and it would be expensive and impracticable for the shipper to select these from stock. Here the agent can play a very useful part in selling the balance of the specification elsewhere, so that the shipper can be saved from uneconomical procedure.

The agent is responsible for keeping his principal informed of market possibilities and should be fully conversant with currency regulations and the money markets of the world. He must also be fully conversant with the shipper's products, thereby being able to advise potential buyers of suitable stocks to satisfy their exact needs.

Having negotiated the sale, the agent will provide the shipper with full delivery instructions, arrange insurance and sometimes, on behalf of the importer, charter a vessel to carry the goods (see Chapter 3, section 2, Contracts). He sees that all shipping documents are complete and that delivery and payment are carried through smoothly, is on hand when the shipment arrives, will investigate complaints and handle claims and is generally responsible for ensuring that the contract is completed to the satisfaction of both parties. Although he does not at any time become the owner of the goods, he often provides credit for both the shipper and the buyer or importer and indemnifies his principal against default by the buyer. For example, he may advance money to the shipper to finance the purchase of logs until shipment is made and payment is claimed. He may also give credit to the importer buying through him. The majority of agents also provide "del credere" facilities, which means that for extra remuneration they will indemnify their employer against loss arising from failure of persons with whom he contracts to carry out their contracts. This additional commission paid by the shipper is in the order of quarter to half of one per cent.

The rate of commission which an agent receives from his principal for his services is entirely a question of negotiation when the agency agreement is drawn up. The rates vary between 2 and 5 per cent and there are no collective agreements to enforce them—it is a matter of individual agreement.

It will be seen that the agent performs a vital function and perhaps his greatest asset is the measure of confidence which he must develop between the two contracting parties. It would be very difficult for a shipper and a potential buyer who may be separated by thousands of miles to negotiate all the necessary functions without having an intermediary who understands the needs of both parties.

Many agents in the softwood and hardwood trades also act as brokers. Whilst there is not much difference, there is, in fact, a distinction, as an agent is directly responsible to his seller principal to the extent to which he has been appointed and authorized and from whom he gets his selling commission. Although not directly material to the argument, the goods the agent sells on behalf of his principal are generally unseen goods and may often be non-existent at the time. He finds buyers for all or part of the goods his principal has to offer but does not himself fix the prices or terms of sale or become responsible in any way for the carrying out of the contract. He may even sign the contract on the seller's behalf but always and only "as agent".

The broker on the other hand deals usually with existing goods as between merchant and merchant—for example, he will find the market for an importer who wishes to dispose of landed stocks or of goods en route to him which his market, for one reason or another, cannot absorb at the time. The goods may be sold to other importers or to non-importing merchants. It may also happen that through unforeseen circumstances, the importer's purchases arrive in this country together and for financial and other reasons he perhaps finds himself incapable of handling them all and therefore employs the broker to remarket some of his purchases.

4. THE IMPORTER

The importer is a merchant who buys timber at first hand from the shipper and this is almost always negotiated through the agent. He then resells either to other merchants or direct to the end-user. There are occasions when a purchase is made direct from the shipper but this is not in accordance with the custom of the trade. Importers do, of course, visit shippers from time to time, maybe to see stocks and to discuss requirements, but even if business is negotiated between them it is still customary for the purchase to be finalized through the shipper's respective agent.

Importers usually take a "general specification", i.e. their purchases from shippers will include a more or less full range of qualities and specification within the class of timber in which they deal. Certain hardwood and softwood importers deal principally with customers in a limited range of industries, and the timber they buy is, in the main, confined to types needed by these industries, much of it coming into the country already cut to the sizes and dimensions and in the qualities which will be required by the ultimate user. In hardwood where the range of species is very wide indeed and imports come from many different countries, importers and agents too tend to specialize in certain types of timber or in those which come from certain areas of the world. Importers' individual purchases of hardwood, whether of logs or sawn timber, tend to be on a smaller scale than those of softwood and plywood, and are often confined to a particular species, though the qualities and sometimes the specifications vary.

The trade of the timber importer gives scope for the exercise of technical skill, knowledge and experience. The importer needs to assess his customer's likely requirements well in advance, to gauge price movements and to familiarize himself with the quality, natural variations and degree of seasoning to be expected of the timber on offer from any particular shipper. In hardwood particularly, the same species may vary in texture from district to district within a single country and the importer needs to know exactly where his timber is obtained. All this is necessary because many months may elapse between the placing of a firm contract and the delivery of the timber.

The importer usually purchases either on f.o.b. or on c.i.f. terms. When purchasing f.o.b. he may make the shipping arrangements for himself but he may instruct the agent or supplier to do this for him (see Chapter 3, section 2, Contracts). In softwood, where consignments are large, it will generally be necessary for him to charter a vessel for a consignment. In hardwood, consignments are smaller and it is generally possible to obtain shipping space for them on vessels carrying mixed cargoes.

The importer usually holds his stocks in a yard or wharf. This may be his own property, or he may rent a yard or space from dock or other authorities, or use public authorities' or public wharfingers' premises for the storage of his goods. Some importers who have their own yards nevertheless from time to time rent additional space according to their needs. All importers have a distributive organization of one kind or another.

Frequently they own their own sawmill where they can convert both the timber imported and the timber bought for conversion by other merchants or by timber users, although some importers rely on public sawmills for this service. The importer may also have his own facilities for drying timber either in the open or in a drying kiln and for treating it in other ways, e.g. with preservatives.

The preceding paragraphs have shown that the buying and selling methods of importers are by no means uniform.

THE TIMBER TRADE

At one end of the scale there is the importer who buys a comprehensive range of specifications, and at the other end the specialist importer who confines his purchases to a more restricted range required only for one or two classes of user. The amount of timber the importer buys already cut and the amount further prepared in his yard also varies from one importer to another, according to the facilities they possess and to the requirements of their customers. In hardwood a good deal of the stock of an importer may be held for anything from 1 year to as much as 5 years before it is sold. Finally, the extent to which importers sell "ex-stock" is by no means uniform and, in normal times, if not at present, the practice of importers differs widely in this respect. Some take the great bulk of their purchases into stock and then sell from their yards; others sell a considerable proportion of their purchases either "ex-ship" or "ex-barge". Some importers do little business outside their own locality, but some of the larger ones have offices and yards in various areas, both at the ports and inland, and sell on a nation-wide basis. A number of importers have interests in overseas shippers' concerns, in agencies or in timber-using businesses. They sell to non-importing merchants and to timber users, and there is also some buying and selling between importers as various needs arise.

5. THE MERCHANT

The merchant should be more correctly described as a non-importing merchant since that is the essential difference between these two sections of the trade. Most importers are also merchants but there are very many merchants who do not import but obtain their stocks from the importers. Merchants are trading in a smaller way and tend to provide particular trades such as building or joinery. It is possible, therefore, that the merchant will find himself competing with an importer for the custom of the end user or actual consumer. He may develop his business to a point where he seeks to dispense with the services of other importing merchants and to become an importer himself.

The merchant often has a decided advantage over the importer from whom he buys his stocks because he has the opportunity to inspect the goods and to buy as near as possible his exact requirements. It is not necessary for him to incur heavy financial outlay nor to carry the large and varied stocks which the importer must have at all times. He performs a very important function by being able to deal efficiently with the day-to-day requirements of the consumer who may have underestimated his requirements or have a specialised or urgent undertaking.

6. THE CONSUMER

The consumer or end-user generally buys his timber from merchants, from both those who import and those who do not. A minority, some of them very substantial buyers, do purchase direct through agents. There are also some special trades, such as that in pencil slats, in the United Kingdom in which it is a long-established custom for the user to negotiate direct with the shipper. In general, however, this is not the custom of the trade.

The bulk of the timber consumed — other than as fuel — is for the following purposes:

Softwood for building and packing-case making.

Hardwood for the furniture industry and interiors of buildings.

Plywood and Fibreboards for the furniture industry and latterly in building on account of the improved durability of the products.

19

It will be seen, therefore, that the ultimate consumer is the builder, case-maker, joiner, furniture manufacturer, etc., who does not often require a wide variety of sizes and qualities and the system as it exists provides for supply to all the varied interests in an efficient and economical method.

3

TRADE PROCEDURE

1. SUPPLY AND DEMAND

As mentioned in the previous chapter, some commercial concerns may produce or deal in several or all of the various sections of the trade. For example, there are firms in Scandinavia who export softwoods, both redwood and whitewood, hardwood in the form of birch, together with plywood and fibreboards. The supply procedure varies slightly within the individual exporting countries and this is influenced by a number of factors such as species, geographical position, climate, etc. If one considers softwood as the predominant import and Scandinavia as a major exporting area, then the following system can be given as an example (Tables II–IV).

The ownership of the forests is shared between private owners, limited companies, public bodies and the State. Each of them is naturally interested in making the best and most of that which they have in the way of income and capital appreciation—the former by sales of logs either standing, felled or delivered and the latter by controlled felling, draining, planting, etc.

The sawmill owners or shippers usually possess some forests of their own but very few of them are self-supporting in that way and have to purchase from other owners in order to maintain their standard of production and quality. Their main source of outside supply is the State Forests, and the annual log auctions are important occasions providing an

3

TABLE II. SUMMARY OF IMPORTS AND EXPORTS OF SOFTWOOD

World softwood imports				
Country	1964 (%)	1966 (%)	1967 (%)	1968 (%)
Belgium—Lux.	1·9	2·0	2·1	1·8
Denmark	2·7	2·3	2·4	2·0
France	3·2	3·3	3·6	3·2
Germany (Western)	9·0	8·2	6·3	6·5
Irish Republic	0·7	0·5	0·7	0·8
Italy	5·6	6·5	7·3	7·0
Netherlands	6·1	5·5	5·5	5·7
Spain	1·6	1·4	1·4	1·4
Switzerland	0·7	0·7	0·6	0·6
United Kingdom	20·5	18·5	19·7	19·6
U.S.A.	25·9	25·9	25·7	27·7
South Africa	0·5	0·3	0·3	0·3
Australia	1·3	1·4	1·4	1·4
Other Countries	20·4	23·4	23·1	22·0
Total	100·0	100·0	100·0	100·0

World softwood exports				
Country	1965 (%)	1966 (%)	1967 (%)	1968 (%)
Austria	6·4	6·4	6·1	6·7
Czechoslovakia	1·7	1·8	1·7	1·3
Finland	9·2	8·7	8·0	8·3
France	0·7	0·7	0·5	0·5
Norway	0·3	0·2	0·1	0·2
Poland	2·0	2·0	1·7	1·5
Sweden	12·0	11·7	12·9	13·7
Yugoslavia	0·6	0·6	0·5	0·5
Portugal	0·9	0·9	0·8	0·8
U.S.S.R.	18·2	18·7	17·4	16·8
Brazil	2·6	2·8	2·4	2·4
Canada	35·1	33·9	34·9	35·7
U.S.A.	4·5	4·8	5·3	5·3
Other Countries	5·7	6·8	7·4	6·3
Total	100·0	100·0	100·0	100·0

TABLE III. PRODUCTION OF SOFTWOOD—SAWN
(thousand cubic metres)

Region	1950	1960	1968
Europe	47,374	55,153	58,029
U.S.S.R.	42,095	89,744	93,750
N. America	88,348	84,797	96,278
C. America	1,542	1,542	2,211
S. America	2,803	3,364	4,452
Africa	42	631	1,102
Asia	13,923	25,532	43,056
Pacific Area	1,588	2,336	2,415
World Total	198,093	263,099	301,293

23

TABLE IV. PRODUCTION OF HARDWOOD—SAWN
(thousand cubic metres)

Region	1950	1960	1968
Europe	8,990	12,850	16,673
U.S.S.R.	7,400	15,840	16,550
N. America	18,790	15,585	17,464
C. America	730	675	941
S. America	2,950	5,475	6,448
Africa	800	1,630	2,050
Asia	5,450	13,330	24,861
Pacific Area	2,430	2,720	2,638
World Total	47,540	68,105	87,625

indication of shippers' ideas of value or a guide to the future of the market. It is not always realized that it may be a year or two before the logs purchased at the auctions have been felled, transported, floated, stored, sawn, seasoned and shipped overseas. During this period market prices can rise or fall quite considerably and these often uncontrollable factors can determine the difference between profit and loss.

It is very important for the sawmiller to be able to forecast the probable requirements of the markets in the various countries to which he is likely to sell, so that he can convert the logs into the appropriately dimensioned sawn goods. At the same time he is limited by what his logs can economically produce. As an illustration of this, it might seem desirable to get the highest cubic yield from logs of a certain diameter by cutting $7\frac{1}{2}$ in. battens. This would not be a true economy when 7 in. is demanded, even if it means increasing the waste from the log. In latter years sawmill waste has gained in value and its disposal is no longer a problem since it is used by the woodpulp and fibreboard concerns in addition to providing fuel and energy.

Each country has its own traditional sizes, and whilst there is much conservatism in the timber trade, there are occasional trends towards a change in required dimension which can hardly be foreseen. In order to anticipate the demand, the sawmiller—who now becomes the shipper and seller—consults his agents in the different importing countries. Once the shipper has been able to estimate what logs he will be getting, what they will economically produce in sawngoods and at what date they will be properly seasoned for shipment, he makes up his stocknotes of quantities, qualities and sizes which will be available for shipment at different dates and sends copies to his selling agents.

The selling agent's function is to find buyers willing to purchase the goods offered by his principal. He must ascertain their requirements and their price ideas and he needs to enjoy the complete confidence of both parties, otherwise he cannot

25

hope to succeed in business. The buyer or importer has to anti-cipate the needs of his customers—usually the merchant—and buy in advance what he expects they will need. He may possibly specialize in certain directions because he has established a reputation for having a continuous supply of certain qualities or marks. Generally speaking, the importer's customer has very varied needs and he will probably invest in a balanced range of imports to cover the raw material requirements. This calls for specialist experience and only those who possess this can hope to profit from the venture. It is not easy to anti-cipate demand in advance and many importers buy forward by as much as 9 months, i.e. a contract is closed 9 months before delivery due to the fact that many Scandinavian ports are frozen until the second quarter of the year.

The negotiation which proceeds prior to the closing of a contract is dependent upon three main factors—specification, price and shipping date—in that order of importance. No business can result unless the buyer can obtain the specification which satisfies his needs, whereas the price may vary according to whether the sizes in question are difficult to supply. When the agent has obtained the importer's offer, he passes this to the shipper—always provided that it is a reasonable price for a reasonable specification. Present-day negotiations are always conducted by telex, telegram or telephone, since delay on the part of any interested party could mean loss of business to competitors. If the offer is accepted, contracts are ex-changed (see section 2, Contracts). Should it be unacceptable on account of specification or price, then the importer must either accept the shipper's counter-offer or renew his offer or improve or withdraw it. The outcome is usually a matter of compromise together with a measure of goodwill, but it should be noted that it is always the shipper who decides what he will sell and at what price—the agent acting as negotiator between the two parties.

The softwood trade has always been a seasonal one, parti-cularly in respect of the Baltic—U.S.S.R., Finland and Sweden.

Ship movement is limited in northern waters for many months in the year due to ice. Consequently one-third of the U.K. imports arrive in the first 6 months of the year, and two-thirds arrive in the last 6 months. However, this pattern has changed somewhat in recent years due to a number of reasons. The increase in kiln drying has meant that timber can be shipped sooner after sawing than was possible when all goods were air seasoned, and modern ice-breakers have helped to keep shipping lanes open longer. There has also been an expansion of all-year-round shipments from Canada to the United Kingdom.

The date of shipment is when the shipper undertakes to have the goods described in the contract ready for loading—seasoned or unseasoned as the case may be (see Chapter 4, Grading of Timber). It is not always easy for a shipper to estimate such a date with accuracy, since so many of the factors already mentioned affect both readiness of the goods and navigational facilities. A very severe winter and late spring can significantly influence seasoning and freedom from ice—known as F.O.W.—first open water.

In regard to hardwoods, this section of the trade is generally more of an all-the-year-round business, since the goods are exported from areas where climate conditions are not so exacting. Because hardwood is usually carried by liners on regular services the shipper is able to secure freight space easily and can therefore offer shipping dates with his stocknotes, which are usually issued the whole year round.

Both shippers and importers in their respective countries have statistics which give a guide towards market trends over the years, and intelligent interpretation of these can help both to anticipate demand. These figures are normally issued by the association which represents them—in the United Kingdom the Timber Trade Federation has sections which cover all sectors of the importing trade. At regular short intervals all members are circulated with figures showing the stock, consumption, purchases and arrivals. This, together with statistics from other sources such as government departments, gives the building figures and therefore reflects the consumption of the softwood trade's best customer. Additionally, world organizations such as F.A.O. produce statistics together with assessments of future trends on a worldwide basis (Fig. 2).

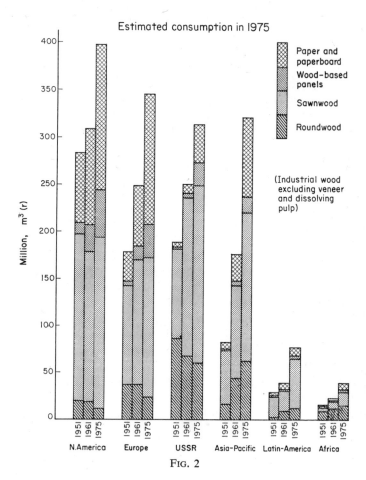

Fig. 2

2. CONTRACTS

General

An agreed sale can be effected by telephone, telegram or in conversation. It is the custom of the trade that such transactions are binding until the formal exchange of signed contracts can be made in due time.

In the United Kingdom the law relating to the sale of goods is set out in the Sale of Goods Act, 1893. Section I of this Act defines a contract of sale as one "whereby the seller transfers, or agrees to transfer, the property in goods to the buyer in return for a money consideration called the price". This statement covers not only the actual sale (which takes place only when specific conditions are fulfilled by each party) but the agreement to sell. It therefore covers the sale of future goods where the seller agrees to transfer the property at some agreed future date. It is important to note that no contract for the sale of goods can have any legal effect unless it is for a "consideration".

All transactions concerning the sale of wood goods are subject to the law of the land. It would be surprising if there were not occasions when differences arise between the parties concerned. All contract forms provide for arbitration—later described in greater detail—and it is usual for any disputes to be amicably resolved in this manner. However, there are sometimes occasions when judicial decisions are necessary and the Courts take the view that, since there is an arbitration clause in contracts, the law of the land in which the contract is closed must apply. In so far as the United Kingdom is concerned, English Law governs the usual timber contracts and this is accepted by both parties because the contract forms were initially agreed by the associations in the respective countries involved. A possible exception to this rule could arise if a buyer visited a seller in his own country and concluded a contract on a form other than the usual printed timber trade contract forms.

29

Contract Forms

Over a considerable number of years, and as a result of negotiations between the exporting and importing organizations of many countries, there have developed a vast number of official printed contract forms. These all contain basic clauses together with special terms and conditions applicable to the countries concerned. In the United Kingdom, the Timber Trade Federation (T.T.F.) has negotiated on behalf of the importing concerns and the official contract forms are given code names usually followed by the year. To mention just a few, the following are current at the time of writing.

Scandinavia:	UNIFORM 1964 (f.o.b.).
Scandinavia:	ALBION 1964 (c.i.f.).
C. Europe:	EUCON 1959.
W. Canada:	PACIF 1958.
E. Canada:	ECANCIF 1958.
Hardwood:	UNICIF 1952.
Plywood:	PLYCIF 1957.
Particle board:	PARCIF 1963.

In the event of the need arising to amend such forms it is usual to retain the code name but change the date. This system is generally adopted by other countries, e.g. the Swedish Wood Exporters Association has contract code names for all the contract forms used to cover sales to countries other than the United Kingdom. These have been agreed with the respective organizations in France, Germany, Holland, Denmark, Spain, etc.

The aforementioned basic clauses printed on the face of a contract provide for the following information to be inserted:

(a) The names and addresses of the buyer, seller and agent.
(b) The name of port of shipment (and ship if known).
(c) The date of shipment.
(d) The name of port of destination.
(e) The specification of goods and prices.

The last item is the most important section of the contract

and covers quantity, description, quality, dimensions, etc. The description of the goods is of vital importance, since it should be remembered that a buyer seldom has opportunity to see the goods as normally a contract is closed for future delivery and possibly before the goods are manufactured. He can only rely on the description to safeguard his interests and is not in the happy position of a consumer who can inspect the goods in the importer's yard.

Section 13 of the Sale of Goods Act states that the sale of goods by description implies that the goods shall comply with that description. Section 14 states that in a contract there is no implied warranty or condition that the goods are fit for any particular purpose. The only implied condition is that the goods shall be of merchantable quality. Therefore, it follows that if the buyer wishes to ensure that he receives goods up to his expected standards, he should be particularly careful to see that the description is complete and that it fully protects his interests.

In addition to this basic information—which obviously varies from contract to contract—there is a second part consisting of standard printed clauses which are known as "the General Conditions". These are usually printed on a separate sheet and are not attached to every contract form as it is generally accepted that the "General Terms, Conditions and Warranties" are known to both parties. This is not to say that certain items cannot be amended, but only by mutual agreement between buyer and seller. These clauses cover the following items: payment, margins for seller and chartering, over- and under-shipment, passing of property, bills of lading, insurance, claims, rejection, arbitration, etc.

Form of Sale

In the interests of simplicity, it can be said that there are two manners in which imported goods are sold. Such contracts are usually referred to as being either c.i.f. (cost, insurance, freight) or f.o.b. (free on board).

31

In a c.i.f. contract, the seller has greater obligation because he is responsible for chartering and arranging insurance cover. He must cover himself against risk of freight rate increases and he generally incurs greater trouble than in a f.o.b. sale—which is why most sellers prefer to negotiate the latter type of contract.

In a f.o.b. contract, the seller is required to deliver the goods to the vessel and the responsibility for chartering and arranging insurance is for the buyer. It should be noted that Scandinavian sellers prefer to use the term f.a.s. (free along-side), which they maintain is more correct since they are only responsible for placing the goods alongside the vessel. This is because the property or ownership of the goods passes when the goods have been placed on board and it is often quite impossible for some sellers to do more than bring a lighter alongside the vessel—there being no shore crane facilities in many ports.

c.i.f. Contract

The broad terms of such a contract are much the same as for a f.o.b. contract except that the shipping documents become of greater significance. The form states that payment must be made against receipt of shipping documents which must include invoice, bills of lading and insurance policy. In other words, it means that these documents must be such that they can be transferred to a buyer so that he is in the same position as if he had made the arrangements himself on a f.o.b. contract.

Once the contract has been signed, the seller's responsibilities are as follows:

(a) To ship the goods described at the named port of shipment.

(b) To arrange freight for delivery to the named destination at the stated date.

(c) To arrange insurance on the goods during passage, this for buyer's benefit.

(d) To provide an invoice at the c.i.f. price.

(e) To send to the buyer the following documents: bills of lading, insurance policy, specification, invoice and other documents where applicable, i.e. certificate of origin, grading certificate, etc. Receipt of these enable the buyer to ascertain that the goods are in accordance with contract, to obtain delivery when the goods arrive and to obtain compensation should goods be lost or damaged during transit.

The buyer's responsibilities are as follows:

(a) To provide the seller with loading orders in good time, i.e. the number of bills of lading and order of stowage.

(b) To pay for the goods on receipt of the documents.

(c) Should the contract be C. & F., only, then the buyer must also arrange for the insurance cover.

f.o.b. Contract

In this form of contract the seller's responsibilities are as follows:

(a) To provide goods of the correct description alongside or on board the vessel in accordance with the buyer's loading instructions.

(b) To advance freight if required.

(c) To provide the buyer with shipping documents consisting of specification, invoice and bills of lading.

(d) To pay any dead freight claims or to provide indemnity (see section 3, Shipping).

The buyer's responsibilities are as follows:

(a) To arrange a freight contract or charter.

(b) To provide loading instructions to the seller in good time.

(c) To arrange insurance for the goods in transit.

(d) To lift the goods by the stated date in the contract.

(e) To pay for the goods on receipt of the documents.

The documents referred to in the preceding paragraphs can be more fully described as follows.

Bill of Lading

This is normally abbreviated to B/L. It is the main "title deed" to the goods and the person who possesses the bill duly endorsed to him is the owner of the goods. It is a receipt by the master of the ship for the goods carried but only in respect of number of pieces, which is why the clause "quality, condition and measure unknown" is included on the document.

Specification

This is a complete tally of the goods and states quality, size and measurement of the whole cargo. It gives a number of pieces of each length and total cubic quantity.

Insurance Policy

When arranged by the seller in a c.i.f. contract, the document must be stamped and endorsed before it is passed to the buyer in order to enable him to claim compensation if necessary when he receives the goods. (See section 4, Insurance.)

Charter Party

This is a contract by which a ship is chartered to carry goods and is often abbreviated to C/P. (See section 3, Shipping.)

Certificate of Origin

This document is provided by the seller in the case of goods imported to the United Kingdom from the Commonwealth countries. Imperial preference can be claimed when passing the documents through the Customs. Such goods are duty free, whereas goods from other countries can be charged with import duty of varying amounts.

3. SHIPPING

General 1650919

Wood goods are carried by two types of ship—cargo vessel and liner. The former is by far the most widely used and the latter is largely confined to hardwoods and plywood.

Cargo vessels are usually "freelance" carriers and carry almost any type of cargo and these are chartered by those responsible through ship brokers. In regard to the carriage of timber, there are certain companies who specialize in this commodity and it is often customary for a ship owner to regularly ply between two countries carrying wood goods one way and other commodities on the return passage.

Liners are vessels which sail on scheduled passages between specific ports and these are not engaged by charter parties since the owners determine the order of events. When freight space is booked to carry, say, hardwoods on a scheduled run from Japan to the United Kingdom, this is confirmed by the owners with a booking note which is a substitute for a charter party.

Regardless of the type of ship engaged for the carriage of wood goods, there are strict regulations which classify as to its condition and seaworthiness. Most of the world's more important vessels are classified by Lloyds Register of Shipping and this is the basis of all British marine insurance policies. This classification is extremely important to the charterer as it affects the insurance cover on the goods being carried. If an unusually cheap freight rate is quoted by a foreign ship owner it is possible that the ship is not classified at Lloyds—in which case the insurance company may not be willing to insure the goods in transit.

All ships are required to keep a ship's log, which is a complete diary of the life and events aboard. This log is often of vital importance and its inspection can be demanded by the bill of lading holder at the port of destination if he has

35

reason to believe that any extraordinary occurrences have arisen during the voyage which may have affected his interests. Reference to the ship's log may confirm bad weather during the voyage and this may have caused damage to or even loss of deck cargo. Shortages in deck cargo can be supported by such information and claims for compensation can be formulated.

Chartering

The initial procedure for arranging freight to carry goods is for the party responsible for chartering—see section 2, Contracts—to contact a ship broker who is known to specialize in the particular area in question. When suitable space is found at a rate of freight which the charterer finds acceptable, then a charter party or freight contract is prepared.

There are two principal types of charter party: (1) a voyage charter and (2) a time charter. A voyage charter covers a vessel engaged for a particular voyage between two particular ports. In the timber trade it is customary for the freight rate to be based on a price per standard with variations for different dimensions, i.e. deals and battens are often different from boards.

A time charter covers a vessel engaged for a definite period of time during which it is under the direction of the charterer. The ship owners are paid so much per week or month irrespective of the amount of cargo which the vessel carries.

In the carriage of wood goods, it is more usual for the voyage charter to be used. Under these conditions the importer who pays the freight is only paying on the actual quantity of goods carried and he has the advantage of knowing his costs in advance. This can be of vital importance if he is selling his goods in advance of arrival since he knows exactly the cost per standard.

Apart from the question of arranging freight at an acceptable rate, there is the problem of estimating the required tonnage or cargo space. It is very easy to over- or under-estimate a ship's capacity and it often happens that the vessel is not able to lift the full quantity of goods for which it was chartered. This can be due to over-estimation on the part of the owners, especially when the vessel has not been regularly used for the carriage of wood goods.

In the event of the charterer not providing sufficient cargo within the margins permitted, then the shipowner will claim "dead freight". This is for the unoccupied space which should have been filled by goods and is payable at the full rate. As a simple example as to how a dead freight claim can arise, consider a contract to lift 300 standards with, say, 10 per cent margin. If the timber is unusually heavy or of high moisture content, then the ship may be down to her marks when only 250 standards have been loaded. The owners will then claim dead freight on 20 standards, i.e. 300 minus 30 equals 270, being the lowest quantity for which they contracted. In such circumstances the owners are entitled to claim that they have lost money due to no fault of theirs. When such claims are acknowledged it is the seller or shipper who is held responsible for payment. At the time of loading the ship's master will "enter a protest", which is done before a notary or a local consul. This document gives all the relevant details substantiating the claim and if the seller or shipper does not agree to its allegations, then he will enter a "counter-protest". Both these documents are forwarded to the agent together with the documents previously mentioned, i.e. bills of lading, etc.

A charter party is a standard form and the many types are often given code names in much the same way as the contract forms previously mentioned. All these contain the following basic details:

(a) Names of the contracting parties—ship owners and charterers.
(b) Details of the ship—name, classification, tonnage, nationality and location at time of chartering.
(c) Anticipated loading date.

37

(d) Description and size of cargo.

(e) Freight rate and how it is to be paid.

(f) Allowable time for loading and discharging and the amount of demurrage due if this time is exceeded.

(g) Names of loading and discharging ports.

(h) Certain clauses which are designed to protect the ship owner's interests in the event of circumstances arising over which he has no control, e.g. the ice clause which covers the freezing of certain Baltic ports and prevents entry of ships and the strikes clause which covers both loading and discharging.

In all the various forms, the undertaking is to "load, carry and deliver" the goods specified in the charter party. There are a number of expressions used in all charters and the following gives a brief description of their meanings.

Demurrage. This is a sum paid or agreed to be paid to the ship owners as damages for delay in loading and/or discharging which is beyond the times stipulated in the charter. Demurrage is usually calculated as a sum per day and it is paid by the charterer.

Custom of the Port. This is not easy to exactly define but it is intended to make allowances for circumstances which are peculiar to some ports—particularly overseas. These are usually well known to ship owners and shippers and include such events as local holidays, action due to adverse weather conditions, local labour stipulations, etc.

Lay Days. These are the days during which the cargo must be loaded and discharged. It is an exact term and is usually referred to as "fixed lay days". This is associated with "weather working days", which is the rate of loading per working day when weather permits without interruption.

Plimsoll Line. This is a white line painted on the hull of a vessel to indicate the safe loading level. This level is fixed by the Board of Trade and is required by the Merchant Shipping Act. Depending on the design of the vessel, it is a level determined by skilled inspectors and it varies slightly for different areas of the world and the seasons. The expression "loaded down to her marks" means that a vessel is loaded to full capacity and this capacity varies with the season—for example, in extraordinarily cold conditions the additional weight of ice on the superstructure has a significant effect on stability.

Deck Cargo. It is usual for the better qualities of timber to be stowed under deck in accordance with the buyer's loading orders. Vessels usually carry approximately one-third of the total cargo of wood goods on deck. Present-day development in modern ship design has increased this proportion and bulk carriers often carry as much as half the cargo on deck. There is considerable significance between deck and under-deck cargo in respect of deterioration and losses and this is described in Chapter 5, Claims and Arbitration.

N.A.A. (*Not Always Afloat*). The clause which includes the name of the port of destination is usually followed by the words "or so near thereunto as she may safely get and deliver the goods always afloat". In some small ports the vessel may have to lie on the mud at low tide and under such known circumstances the words "always afloat" are deleted and substituted by "N.A.A.".

Lien. The vessel's master or owners have complete lien on a cargo for all charges arising such as freight, dead freight and sometimes demurrage. This means that if the owners do not receive payment of such items, they can hold the cargo and refuse to deliver it until they receive satisfaction.

Advance Freight. It is customary for provision to be made available to the vessel to cover incidental expenses at the loading port. This does not exceed one-third of the total freight and the amount is endorsed on the bills of lading which leaves a balance of freight when the vessel reaches port of destination.

39

4. INSURANCE

General

Marine insurance is effected in the form of a contract between two parties, the insurer and the assured. The former undertakes to provide cover against loss through maritime perils for the latter in consideration of a specified payment called a premium. Since it is a contract of indemnity, it is only possible to recover the amount of loss actually sustained.

The assured must have an insurable interest and this may be in respect of the following: (a) the vessel's hull, (b) the freight or (c) the cargo. The ship owners have an interest in the hull and the importer has an interest in the cargo, freight and imaginary profit. The transaction is usually effected by insurance brokers following approach by the importer or shipper—whichever is responsible according to the form of contract. The broker draws up a suitable contract, which is called a policy, and the insurance cover is placed with an insurance company or underwriters. Alternatively, the insurance cover can be placed direct with an insurance company without the services of a broker.

The most famous name in marine insurance is Lloyds of London. This organization can be described as an exchange of which underwriters are members and it is customary for a group or syndicate of them to cover the risks in question, i.e., cover is not the financial risk of any one individual underwriter. A contract or policy is itself covered by the Marine Insurance Act of 1906, which states that the following details must be shown:

(a) The name of the insurer and the assurer.
(b) The subject of the insurance, i.e. wood goods, etc.
(c) The risks against which the subject is insured.
(d) Details of the voyage—naming the ports concerned.
(e) The sum insured, which, in the timber trade, is usually the f.o.b. value plus 10 per cent imaginary profit plus prepaid or freight advance as applicable.

There are a number of different kinds of policies but in so far as the timber trade is concerned, the most usual type employed is a "voyage policy". This covers goods from one place to another during one particular voyage. A standard type of policy form is used as it would be clearly impracticable to draw up a fresh one for each voyage. The customary standard type contains clauses which have existed for very many years—long before the Marine Insurance Act, 1906—and in the light of present-day conditions refer to some perils which no longer exist, such as piracy. It remains unchanged because over the years there is no ambiguity in any section—all of which have been tested and interpreted by the legal world. If a new form was produced with modern clauses, then it is possible that the legal meaning would have to be reinterpreted and court rulings obtained when disputes arose. Any anomolies are overcome by the insertion of special clauses to suit the class of goods which are covered.

Losses Covered

In a voyage policy, the cover commences from the time when the goods are loaded onto the vessel. The goods are not covered prior to loading—except by the special additional T.T.F. insurance clauses which are explained later. The insurance cover ceases when the goods are safely discharged at the port of destination.

The principal losses which are normally recoverable under a marine insurance policy are as follows:

(a) Total loss—absolute or actual.
(b) Partial loss or damage.
(c) General average.
(d) Salvage loss or charges.

(a) Absolute or actual total loss covers not only complete destruction but also damage which is sufficiently severe to render the goods useless or where the assured is permanently deprived of the goods in question.

41

4*

(b) Partial loss or damage is known as "particular average". It is any partial loss or damage caused by any of the perils insured against. There are certain types of cargo, such as wood goods, which are more likely to suffer damage than others. For example, deck cargo is easily damaged by seawater and such deterioration is usually excluded by underwriters unless there is a major peril to the vessel such as fire or collision. The exclusion of these risks is known as "free from particular average" and is abbreviated to f.p.a.

(c) General average can be described as an obligation on all those who have an interest in the vessel and its cargo. It is the voluntary sacrifice of the minority in favour of the majority interest and this situation can arise if it is found necessary to jettison part of a deck cargo to save a vessel under severe weather conditions or if it runs aground. Under such conditions, the sacrifice of part of the cargo could mean the saving of the whole cargo. The value of the part sacrificed is equally shared by all the bill of lading holders and the ship owners. This "contribution" is assessed in accordance with the York–Antwerp rules of 1950. These individual contributions are themselves covered by insurance.

(d) Salvage losses are those arising from the insured perils leading to the necessary selling of the goods before they reach their destination. Salvage charges are those incurred in order to save the vessel in peril.

T.T.F. Insurance Clauses

The standard form of policy already mentioned is phrased in rather general terms and covers the goods as and when they are placed on board the vessel. Furthermore, the cover is restricted to a well-defined voyage. The Timber Trade Federation of the United Kingdom in agreement with the appropriate insurance associations produced special clauses which were applicable to the special requirements of the timber trade. These are not only accepted by all reputable insurance companies but also by all the important exporting countries of the world. Policies which bear the words "Including Timber Trade Federation Clauses" are extended to cover all the additional risks and contingencies.

The most important of these additional clauses is the transit clause—generally known as the "warehouse to warehouse" clause. This extends the cover right through from the shipper's yard to the final destination—both of which may be some distance inland from the ports. The phrase "in transit" gives wide cover since it includes ordinary trans-shipment, delays in transit warehouses, etc. However, it does only apply provided that the goods are truly in transit and that the journey is not broken in any way—such as storage at a port for the convenience of the assured.

The risks covered are intended to mean all physical loss or damage from fortuitous circumstances. Claims arising from inherent vice or delay are not covered. This latter exclusion is of primary importance since delay could cause deterioration of a cargo of wood goods, say, in the form of discoloration should the moisture content have been too high for shipment to the country in question. The assured would have a legitimate claim if the deterioration was found to be due to flooding of the hold, contamination by other goods, etc., i.e. not due to inherent vice in the wood. Other clauses refer to increased values, war and strike risks and collision circumstances.

GRADING OF TIMBER

1. GENERAL

The segregation of individual pieces of timber in terms of quality and condition is known as grading, sorting or bracking—the latter term being customary in Scandinavia. It is evident that all exporting mills aim to gain the highest possible yield of best-quality timber since this gains the highest prices. At the same time, the shipper has to consider the maintenance of his reputation and to ensure that his standards do not fall below those set by previous years or by the rules laid down by the authorities in question. For example, should the mill be unfortunate enough to receive a very poor run of logs from an inferior forest area, then the correct procedure is to maintain standards and accept the low yield of high-quality timber.

The principles of commercial grading are different for softwood and hardwood. Softwoods—whether from the Baltic countries, Canada or Central Europe—are normally used for construction purposes in the sizes in which they are shipped and the grading is based on that assumption, i.e. the extent of the permitted defects depends entirely on the dimension. An 18-ft length of 2 in. by 6 in. may have a cluster of knots in the middle section which renders it unfit for construction use despite the fact that maybe 90 per cent of its length is otherwise faultless. Alternatively, if these defects were in the end section, then the piece would be reduced in length and upgraded—possibly finishing as a 16-ft length.

Hardwoods are graded for their "cutting" and not "construction" value. They are evaluated according to the amount of wood free from defects and of reasonable size which can be cut from a piece. The existence of a serious defect which can be cut out, leaving the rest of the piece faultless, entitles the whole to qualify as high grade. One can readily see the reason for this—hardwoods are rarely used for anything other than furniture and interior decoration and appearance is all-important, together with utilization of relatively short lengths.

All grading is a visual process and, although done by graders of considerable experience, it follows that human error can be involved. It would be extremely surprising if differences of opinion did not exist and for that reason it is generally conceded that even the most reputable shipper will occasionally include some doubtful pieces in a cargo. It is equally true to say that such mistakes work both ways and one can often find pieces which are too superior for the grade in question. Therefore, it is both fair and reasonable to view a parcel of wood goods as a whole in order to gain an impression of its quality.

Associated with quality is the very important question of condition. This refers to the state of the timber and covers freshness, weathering and discoloration. A good-quality parcel of timber can be ruined by being shipped in bad condition—such as being insufficiently seasoned. However efficient the actual grading in respect of defects may have been, such circumstances entail degrade. This subject is covered in greater detail in Chapter 5. At this point, it is also advisable to mention the difference between the procedure adopted by the West Coast Canadian shippers and all other softwood-exporting countries. All West Coast Canadian timber is shipped in a green or unseasoned condition, whereas softwood from all other sources is normally seasoned to a shipping dry condition—around 20–24 per cent moisture content.

2. NORMAL METHOD GRADING

It is customary for grading at the mill to commence in the log pond where the logs are kept just prior to sawing. Apart from segregation into the various diameters to suit the setting of the saws, in many mills the logs are also sorted by quality. This can only be an approximate operation since close examination is difficult, but it is possible to divert the obviously low-quality logs.

After sawing, the pieces pass directly to a moving conveyor which is known as the "green chain"—the wood being green or unseasoned at this stage. This is where the initial sorting is done and it must be emphasized that with the exception of West Coast Canadian timber it is by no means the final sorting—this latter operation is done by all other softwood-exporting countries after seasoning. The main purpose of the green chain sorting is to segregate the various dimensions into approximate quality units so that each seasoning pile or kiln load will contain pieces of equal dimension and very nearly equal quality. Since the green timber is moving at a steady pace along the green chain it follows that the grader can do no more than quickly assess its quality—there is no time to count and measure knots or other defects. Such men are of wide experience and it is seldom that they fail to notice abnormal defects. Each piece is marked with the grader's mark and as it passes on it is pulled off the chain onto its appropriate bogie—which in due course conveys the load to the seasoning area. At this point it is customary for softwood to be anti-stain treated to inhibit discoloration development or sap stain. This usually consists of dipping or spraying with a chemical such as sodium pentachlorophenol.

The final grading, and the one of greater significance, is done after seasoning. This is because certain defects such as splits and shakes, distortion and discoloration, may have developed which may necessitate reduction to a grade lower than was indicated by the initial green chain grading. The

46

actual method of seasoning is at shipper's option and the present-day tendency is towards kiln-drying, since this is more economical compared with months of air seasoning. It should be noted, however, that this does not mean more than shipping dry moisture content, i.e. it is not necessarily low enough for end use.

When seasoned, the timber is crosscut to an accurate length, graded and marked on the end with the appropriate shipping mark. Prior to this crosscutting the ends of the pieces have been roughly cut and it is usually necessary to remove splits and shakes which develop at the end during seasoning. It is customary for shippers to have different marks for each of the grades as a form of identification. With the exception of Russian timber—which bears a hammer mark on the end—all other softwood is usually marked with letters in red paint.

3. GRADING RULES

Most major exporting countries have official grading rules which have been produced after consultation by the appropriate authorities representing the sawmill industries.

Softwoods

Canada

The principal West Coast softwoods are Douglas Fir, Hemlock (including Balsam), Western Red Cedar and Sitka Spruce. These are covered by grading rules known as the Export "R List" and these were published in 1951 by the Pacific Lumber Bureau and adopted by the West Coast Lumberman's Association (U.S.A.) and the British Columbia Lumber Manufacturers Association (Canada). All exported timber is graded in accordance with these rules; this is specifically mentioned in contracts of sale and an inspection certificate is provided with the shipping documents.

Eastern Canadian Quebec Spruce is covered by grading

rules drawn up by the Maritime Lumber Bureau and these are known as the M.L.B. Grading Rules for Eastern Canadian Spruce.

Finland

In 1961, the Association of Finnish Sawmillmen produced revised grading rules to cover the two principal exported softwoods—redwood and whitewood. These revised rules— like those of the previous 1936 ones—are "intended as a general guidance for grading of sawngoods for export and do not constitute any binding rules for the individual mills". This statement, which is quoted from the preface to the rules, clearly indicates that the guidance provided is intended to unify standards and does not necessarily mean that all exported timber is strictly in accordance with the rules. It is customary to classify Finnish exports from individual mills as "shipper's usual" and in fact the two official contract forms— Uniform and Albion—do specify "each item to be of Shipper's usual bracking, average length and fair specification for such description of goods". No mention is made of the grading rules.

Sweden

Also in 1961, the Association of Swedish Sawmillmen produced revised grading rules which were drawn up by a special Timber Grading Committee. This committee worked closely with the Finnish committee and the rules were published simultaneously in both countries. Since both rules were in accord on all essential points, the foregoing remarks for Finland also apply for Sweden. Both rules are comprehensive and contain colour photographs to illustrate defects in the various grades.

Soviet Union

No officially published grading rules exist but all redwood is of Leningrad bracking and all whitewood is of White Sea bracking. Since the State is in effect the sole exporter of wood goods, there are standard shipping marks (hammer only) to indicate the regional sources of supply—the more important being as in Table V.

TABLE V

	Unsorted	Fourths
White Sea Ports		
Archangel	S u A	S IV A
Archangel	S u/sA	S IV A
Onega	S u O	S IV O
Mesane	S u M	S IV M
Petchora	S u P	S IV P
Kem	K u K	K IV K
Kovda	K u KD	K 4 KD
Keret	K u KR	K 4 KR
Oumba	K u U	K 4 U
Kara Sea Ports		
Igarka	KS u EI	KS IV EI
Igarka	KS u E	KS IV E
Baltic Port		
Leningrad	E L	E 4 L

United Kingdom

The use of softwood for constructional purposes in the United Kingdom is, in broad terms, covered by two British Standard specifications:

(1) B.S.1860—Structural Softwood—Measurement of Characteristics Affecting Strength.

(2) British Standard Code of Practice C.P.112—The Structural Use of Timber in Buildings.

Recommendations in the Code of Practice are based on the performance of different species in laboratory tests and the effect of strength-reducing defects such as knots, shakes, slope of grain, etc. Since it is the size and number of these defects which ultimately determine the strength of a piece or of a structure, a schedule of maximum sizes is included in the Code.

49

Hardwoods

The most widely accepted hardwood grading rules are those published by the National Hardwood Lumber Association (U.S.A.) and usually referred to as the N.H.L.A. Rules. These are very comprehensive and clearly defined and for this reason have been adopted by many other hardwood-exporting countries outside the United States. These rules cover many specialized grades together with standard grades for different species.

The Malayan Department of Forestry has also produced comprehensive grading rules known as *Malayan Grading Rules (Export) for Rough Sawn Timber*. These are also comprehensive and are generally regarded as being the counterpart of those produced by the United States.

4. GRADES

Most exporting countries have their own particular grades which are most suited to their raw material and production facilities. Therefore, it is desirable that these should be described under a country of origin heading.

Canada (West Coast)

The Export Grading Rules provide strict limits for the various grades of Clears, Merchantable and Common. Certain tolerances and variations are permitted, not only in quality but also in dimension. In the general notes, the following significant statements qualify the application of the rules.

> The inspection of timber is the analysis of the quality of the product. Inflexible rules for the inspection of timber are impossible, therefore variations determined by practical experience must be allowed. The analysis being visual, mathematical precision is impossible, and therefore a reasonable difference of opinion between inspectors must be recognized.
>
> The grade of timber, as determined by the inspector, applies to size, form, condition or seasoning at time of original inspection.

50

Any subsequent change in the timber must be disregarded in determining the accuracy of original inspection.

Suitability for construction purposes, in the shapes and sizes in which timber is ordered and shipped, will be taken into consideration in grading material according to the grades contained herein. It is not intended to supply material guaranteed to be suitable for remanufacturing into smaller sizes.

Unlike most other exporting countries' procedures, West Coast goods are shipped green or unseasoned—but usually anti-stain treated—and the rules permit shrinkage during subsequent seasoning and variation in sawing. The various grades are designed to provide a range to satisfy all reasonable constructional requirements. The preamble to each grade describes its suitability as follows:

No. 2. Clear and Better (*Clears*). Shall be sound timber, well manufactured.

Selected Merchantable. Shall be sound, strong timber, well manufactured and suitable for high-class constructional purposes.

No. 1 Merchantable. Shall be sound strong, timber, well manufactured and suitable for good, substantial constructional purposes.

No. 2 Merchantable. Shall be sound timber suitable for ordinary constructional purposes without waste.

No. 3 Common. Suitable for general utility purposes.

Each of these grades is further qualified by thickness—the defects being proportional to dimension. These are defined in the following manner and one grade is given here as an example.

Douglas Fir—Selected Merchantable. (Applying to sizes 6/4 and thicker, but not including 3.)

Shall be sound strong timber, well manufactured and suitable for high class constructional purposes. Must be Close Grain, i.e. an average on either one end or the other of the piece of not less than 6 annual rings per inch measured over a 3 in. line measured at right angles to the rings.

51

Will admit the following or their equivalent:

Knots—sound and tight, ranging from approx. $1\frac{1}{4}$ in. in 4 in. widths to $2\frac{1}{2}$ in. in 12 in. widths. Proportionate in wider widths.

Spike Knots—equivalent.

Pitch Pockets and/or Pitch Blisters—medium.

Pitch—scattered streaks.

Sap—$\frac{1}{2}$ width or equivalent (other than black).

Split—approximately $\frac{1}{2}$ width of piece.

Wane—slight.

Wormholes—pin, occasional, scattered, in K.D. stock only.

Variation in Sawing—occasional, slight.

In addition, there are intermediate grades which cover a number of special requirements such as flooring, ceiling, door stock, pipe stock, ship decking, stagings and railroad ties and crossings.

Canada (East Coast)

Unlike the West Coast practice, Eastern Canadian Spruce is normally exported in shipping dry condition. The rules cover grades from 1 to 6, together with the more recently introduced M.L.B. Merch. Grade. The description under which goods are exported varies between the different shippers, but in broad terms the Unsorted consists of Firsts to Fourths and a Fifth quality or Firsts to Thirds and a Fourth quality. Alternatively, the goods may be described as Grade III and Better or Grade IV and Better. The Merch. Grade consists of Selected IVths and Better or IIIrds and Better with a few selected IVths.

Finland and Sweden

As mentioned earlier, the principle employed in grading is very similar in these two countries. Each exports three grades:

(1) Unsorted—the normal grade for joinery and first-class work. Consists of Firsts, Seconds, Thirds and Fourths.

(2) Fifths—for carcassing, constructional work and low-class joinery.

(3) Sixths (Utskott)—for lower-class carcassing work, packing case manufacture and underground or temporary work.

Grading is based on the defects present in both redwood and whitewood and can be classified in two main groups as follows:

(A) DEFECTS IN QUALITY

(1) Structure—knots, pitch pockets, bark pockets, compression wood, rot, insect damage and cross grain.

(2) Manufacture—wane, incorrect measure, imperfect surfaces and bad sawing.

(3) Shakes—those inherent in the tree and seasoning checks and splits.

(4) Deformities—distortion in the form of bowing, twisting, springing and cupping.

(B) DEFECTS IN CONDITION

(1) Moisture—should not be above "shipping dry", which means seasoned to a degree of dryness to stand normal sea and land transit to country of destination—normally 20 per cent with a maximum of 24 per cent.

(2) Blue—discoloration confined to the sapwood and although caused by fungi, is not wood destroying. Log blue develops in the log before sawing and deal yard blue develops after sawing and is mainly confined to the surface.

(3) Other discoloration—changes in surface colour which takes place during seasoning and storage—weathering and browning.

As a general rule, grading is based on the better face and the edges of the piece. The dimension of the piece generally decides the size and number of the permitted defects. As with Canadian practice, it follows that the goods are graded according to the exported size and recutting is not considered.

In very broad terms, the following maximum defects are
permitted:

Firsts

Knots—per 5 ft length—sound, one $\frac{3}{4}$ in. plus smaller
scattered sound and dead pin knots.

Shakes—on face—depth 10 per cent of thickness, length
20 per cent of piece. Reverse face free. On edge—50 per
cent of the face.

Blue—none.

Wane—on one edge—15 per cent of thickness and length
of piece. On two edges combined—20 per cent of thickness
and length.

Insect damage—none.

Rot—none.

Compression wood—none.

Pitch pockets—one allowed—small, shallow and closed.

Bark pockets—none.

Seconds

Knots—per 5 ft length—sound, one $1\frac{1}{4}$ in. plus one $\frac{2}{3}$ in
dia. and scattered pin knots.

Shakes—on face—depth 20 per cent of thickness, length
30 per cent on one face or 40 per cent of combined faces.
On edge—50 per cent of the face.

Blue—none.

Wane—on one edge, 20 per cent of thickness and length
of piece. On two edges combined—25 per cent of thickness
and length.

Insect damage—none.

Rot—none.

Compression wood—permitted if no distortion results.

Pitch pockets—permitted to slight extent.

Bark pockets—none.

Thirds

Knots—per 5 ft length—sound, two $1\frac{3}{4}$ in. plus two $\frac{2}{3}$ in dia. and smaller scattered sound and dead knots and normally occurring pin knots.

Shakes—on face—depth 30 per cent of thickness, length 50 per cent on one face or 70 per cent of combined faces. On edge—50 per cent of the face.

Blue—none.

Wane—on one edge—25 per cent of thickness and length of piece. On two edges combined—35 per cent of thickness and 30 per cent of length.

Insect damage—none.

Rot—none.

Compression wood—permitted if no distortion results.

Pitch pockets—permitted if small and not deep.

Bark pockets—permitted if small and shallow.

Fourths

Knots—per 5 ft length—sound, two $2\frac{1}{4}$ in. plus three $\frac{2}{3}$ in dia. and smaller scattered sound and dead knots and normally occurring pin knots.

Shakes—on face—depth 40 per cent of thickness, length 65 per cent on one face or 90 per cent of combined faces. On edge—50 per cent of the face.

Blue—permitted to very slight and superficial degree.

Wane—on one edge—30 per cent of thickness and length of piece. On two edges combined—50 per cent of thickness and 40 per cent of length.

Insect damage—none.

Rot—none.

Compression wood—permitted provided it does not seriously affect the shape of the piece.

Pitch pockets—permitted if small and not penetrating.

Bark pockets—permitted if small and not penetrating.

55

Fifths

Knots—per 5 ft length—sound, two 3 in. plus four $\frac{2}{3}$ in. dia. and smaller sound and dead knots and normally occurring pin knots.

Shakes—permitted over the whole length of the piece, partly right through, provided it holds well together.

Blue—permitted but should not be deeply penetrating nor too widespread over the whole piece.

Wane—on one edge—50 per cent of thickness and 60 per cent of length of piece. On two edges combined—70 per cent of thickness and 60 per cent of length.

Insect damage—permitted to very slight extent in a few pieces.

Rot—small area of hard rot permitted but not penetrating.

Compression wood—permitted.

Pitch pockets—permitted.

Bark pockets—permitted provided they do not penetrate through the piece.

Sixths

Knots—permitted without limit of number and size provided piece holds well together.

Shakes—permitted practically without limit provided piece holds well together.

Blue—permitted.

Wane—permitted practically without limit.

Insect damage—permitted in maximum of 30 per cent of the number of pieces but not so that the greater part of any one piece is affected.

Rot—hard rot permitted and even penetrating through the piece. Soft rot permitted to a limited degree.

Compression wood—permitted.

Pitch pockets—permitted.

Bark pockets—permitted provided it does not form an open hole through the piece.

It should be noted that these defects are the maximum permitted in the grade, which must also contain a fair proportion of high-grade pieces. Additionally, two major defects do not usually occur in one piece simultaneously.

Soviet Union

The principles employed in grading redwood and white-wood are similar to those of Finland and Sweden. The essential difference is the description of the exported commercial qualities. Russian Unsorted consists of Firsts, Seconds and Thirds and a separate grade of Fourths as the general carcassing or constructional grade—more or less corresponding to Scandinavian Fifths.

As with the exports from Finland and Sweden, the quality is consistent with latitude, which means that the superior goods are those originating from northern areas where the trees are slow grown and of close texture. Therefore, it follows that, in terms of superiority, one can list the areas as follows: Kara Sea, White Sea and Leningrad.

Shippers' Marks

The marking of timber is associated with grading and its significance should be understood. Timber is purchased on description and the mark is not only the producer's trade mark but it also indicates the grade. This applies to softwoods, hardwoods, plywood and board materials.

Such information is contained in a world-known book called *Shipping Marks on Timber*. This gives all the information required to identify marks, shipper, grades and ports of shipment—the latter being a guide to the areas from which the raw material originated and therefore its quality. Examples from the book are reproduced here as Tables VI and VII.

57

TABLE VI. FINLAND: EUROPEAN SHIPPERS AND THEIR MARKS

Name of shipper	Description	Average annual production in standards	I	II	III	IV	V	Unsorted
Metsäliitto Osuuskunta, Helsinki								
Metsä-Saimaa production								
Ship from Kotka and Hamina	Sawn and planed	10,000					SMA	SAIMA
Äänekoski production	Sawn and planed	8000					Ä—E	ÄÄNE
Ship from Mäntyluoto								
Metsäkeskus production	Sawn and planed	5000					K—M	White MEM Red M ⊕ M
Ship from Kotka, Yxpila and Rahja								
Other productions	Sawn and planed	1000					M—A	METSA
Ship from Toppila and Röytä								
Multian Saha, Nuoranne & Pojat, Multia	Sawn						M—I	MULTI
Ship from Mäntyluoto								
Närpes Såg Ab., Närpiön Saha O/Y., Närpes	Sawn goods	4000					N—K	N ⊕ K
Ship at Kaskö								

						VI N—A	I—N—S	I—N—⊛ S
Nikkisen Saha, Mäntsälä Ship from Sörnäs	Sawn							
Nisulan Saha, Rutalahti Ship from Kotka and Hamina	Sawn							NI—SA
Nokia A/B., Nokia Ship from Rauma and Mäntyluoto	Sawn goods	4000					NOKI	NOKIA

(Courtesy of Benn Brothers Ltd.)

59

TABLE VII. MALAYSIA: COUNTRIES OUTSIDE EUROPE—SHIPPERS AND MARKS

Name of shipper	Description	Average annual production	Marks	Quality
Leonowens (Malaya), Ltd., Louis T., London, Singapore and Kuala Lumpur Ship from Singapore, Port Swettenham and Penang	Sawn lumber and logs		LTL Black and white ends; also green and white ends	All grades
Lindeteves-Jacoberg (Far East), Ltd., Singapore and Kuala Lumpur Ship from Singapore and Port Swettenham	All Malayan hardwoods, sawn and planed		JVDB Brown/red Purple/brown/black	All qualities
Lisoon Trading Co., Kuala Lumpur Ship from Port Swettenham	Keruing, light red meranti, dark red meranti, jelutong, etc.	Over 6000 tons of 50 cu. ft		
Lyton Corporation Ltd., Kuala Lumpur Ship from Malayan ports	Sawn		LYTON Red/white/blue	All grades in accordance with Malayan Grading Rules

Ng Trading Co., Selangor Ship from Port Swettenham	D.R.M., I.R.M., keruing, mersawa, damar minyak, jelutong, etc.	10,000 tons	Green/white/grey	Select and better grade and standard and better grade
North Borneo Timbers Sdn. Berhad., Sandakan Ship from Sandakan and Wallace Bay	Round logs	10,000,000 cu. ft		Logs: Prime, seconds, F.A.Q., S.S.Q., S.Q. According to Sabah Standard Grading Rules Ply grade—Normal "trade" grading

(Courtesy of Benn Brothers Ltd.)

5

CLAIMS AND ARBITRATION

1. GENERAL

When one considers the vast quantity of wood goods which change hands throughout the world together with the many qualities and dimensions and the number of hazards involved in transit, it must be conceded that it is inevitable that occasional difficulties will arise and that disputes will require settlement. Both importers and exporters would be optimistic if they did not recognize these possibilities and, for this reason, all contracts do contain a clause outlining the procedure for formulating claims.

The most common form of claim is formulated by the importer or receiver of the goods and this can be directed towards three different parties depending upon the cause of complaint—(1) the shipper or seller, (2) the insurance company or (3) the ship owner.

(1) The seller is responsible for claims involving quality, condition (when shipped), and quantity supplied against contract stipulation and measurement. He may also be responsible for any breach of contract warranty involving damages.

(2) The insurance company is responsible for loss of and damage to a cargo during the voyage—excluding deck cargo damage by, say, sea water. He may be responsible for loss of deck cargo caused by rough weather.

(3) The ship owner is responsible for shortages, i.e. the difference between bill of lading quantity of pieces and those ultimately delivered. (The ship's master signs the bill of lading

for number of pieces only and is required to deliver that number on discharge.)

It should be noted that in every instance responsibility has to be proved and the negotiation for the settlement of claims is normally effected by the agent on behalf of the parties concerned. It is very seldom that amicable settlement cannot be arranged, but when this does fail the matter is referred to arbitration—see section 3, Arbitration.

2. FORMULATION PROCEDURE

The shipping documents should be received in advance of the arrival of the goods and these should be carefully examined for correctness. If these do not agree with the contract conditions or are incomplete, then they should be returned to the agent. If the documents appear to be in order, then they should be checked to determine that the following are correct:

(a) Quantities—over- or under-shipment.
(b) Dimension and average length—if latter is stipulated in the contract.
(c) Deck cargo in accordance with loading instructions.
(d) Invoice prices in accordance with contract.

It is at this stage, should the documents reveal that contract conditions have been unfulfilled, that the receiver can reject the goods. If he accepts the documents and pays for the goods, then he can only recover any loss by claiming against the seller. There may be faults which can only be determined when the actual goods are examined, i.e. quality, condition and shortages. It follows that examination of the documents is of primary importance.

Actual examination of the goods should be carried out immediately upon arrival. Contract conditions stipulate a time limit within which claims must be formulated and, as these are of a relatively short period, it is essential that prompt action should be taken to ensure that one is not "out of time". The Baltic contracts provide for a period of 14 days from date of ship's final discharge whereas the Canadian and certain Central European contracts allow a period of 21 days. Within this period, formal notice of claim, together with "reasonable particulars", should be submitted to the agent. The Uniform—Baltic f.o.b.—contract states:

> No claim for quality and/or condition will be recognized by Sellers upon any goods shipped under the contract, unless reasonable particulars are given to Agents within fourteen working days from date of vessel's final discharge. Reasonable particulars shall mean a statement as to whether the claim is for quality and/or condition together with a statement of the sizes complained of and an estimate of percentages and of the amount claimed. All such statements are without prejudice and conditional on the facilities for inspection.

This clause is typical of most contract claims and clearly implies that it is not sufficient to merely give notice of claim unless it is accompanied by reasonable particulars. It can happen that, due to congestion in the docks or delays beyond the importer's control, it is impossible to comply with this time limit. In such circumstances, the correct procedure is for the agent to be informed and the seller's authority obtained for an extension of the claim period. This is usually granted but may exclude subsequent claims in respect of condition or deterioration brought about by adverse weather conditions over which the seller has no control. These clauses are intended to give protection to the buyer without unfairness to the seller.

On arrival of the goods, the receiver should satisfy himself on the following points:

(a) Condition of the goods. If signs of a rough passage are seen, then reference to the ship's log is desirable. This will indicate shifting or loss of deck cargo, flooding of holds and any other unusual occurrence during the voyage. Condition refers to the state of timber such as discoloration and freshness—the latter not to be confused with wetness due to subsequent wet conditions. Such deterioration can be present at the time of shipment, in which case it is the shipper's responsibility, whereas subsequent wetting causing "sweating" in the hold during an abnormally long and delayed voyage could be the insurance company's responsibility. In many instances, it is difficult to decide where responsibility lies and, under such circumstances, it is usual for the receiver to notify both the seller's agents and the insurance company.

(b) Quality of the goods. Without doubt, this is the most difficult point of all to determine. It requires considerable experience to assess the quality of a parcel of wood goods whether or not they are sold as complying to recognized grading rules or as "shipper's usual". The quality of timber is judged by the presence of inherent defects such as knots, wane and rate of growth and the highest grade can be ruined by being shipped in bad condition or by subjection to adverse conditions during a voyage. If the receiver has any doubts concerning quality he should seek the experience of the agent, who will make representation to the seller. It should be noted that if a claim is for "quality", then the receiver must not "break bulk". This means that the item complained of must be kept intact so that, in the event of arbitration, it can be produced in its entirety as discharged. If the claim is for "condition", then the receiver can dispose of any part without detriment to his claim on the balance of the parcel.

(c) Dimensions of the goods. The thickness, width and length of the goods should be checked. As a very general rule it is customary for all wood goods to be of full measure after seasoning, with the exception of West Coast Canadian softwood, which is often "scant". The grading rules permit an allowance for shrinkage during drying and for variation in sawing. If justified, such claims are always submitted to the seller.

(d) Pieces and total measurement of the goods. As already mentioned, the ship signs for number of pieces only and is responsible for discharging that number. If the discharge tally shows a shortage, then the receiver submits his claim in one of two directions. The insurance company is responsible if the deck cargo is lost and the ship owner is responsible for the under-deck cargo. If the dimensions and pieces are correct, then it follows that the total volumetric quantity is also correct.

It should be noted that if the claim is against the seller, then the inspection is made by the agent who represents that seller. If the claim is against the insurance company, then the inspection is made by an appointed surveyor or claims assessor. In instances where responsibility is in doubt, then a joint inspection is often made. In any event, it is always desirable that there should be a minimum of delay. When condition is involved, circumstances which influence this can change so quickly and the goods should be seen in the "arrived condition".

3. ARBITRATION

Reference to arbitration results from inability to effect an amicable and satisfactory settlement of a claim or dispute between the parties concerned. It is governed by the Arbitration Act, 1950, and all the official contract forms contain a clause which modifies this to conform to timber trade procedure. It can be regarded as the ultimate safeguard for both the parties of the contract. Considering the vast volume of business which is involved there are very few arbitrations in the timber trade. A contributory factor to this is the fact that reference to arbitration can be a costly business for one or both parties, whereas amicable settlement costs nothing—seller's agents performing the task without fee. The agent will always do his utmost to effect a fair and reasonable settlement since inability to do so could be regarded as failure on his part to enjoy the confidence of both parties—particularly that of his principal, the seller.

Before any arbitration, there must be agreement between the two involved parties for such submission. The Uniform and Albion contracts state:

> Any dispute and/or claim regarding shipped goods which it may be found impossible to settle amicably shall be referred to arbitration to be held in the country of destination of the goods.

66

The reference shall be to a sole Arbitrator mutually agreed upon. In default of such agreement claims of more than £100 shall be referred to two Arbitrators, one to be appointed by each party.

Should the Arbitrators fail to agree upon an award they shall appoint an Umpire.

An award shall be final and binding upon both parties. The costs of such arbitration shall be left to the discretion of the Arbitrator(s) or Umpire. In deciding as to costs the Arbitrator(s) or Umpire shall take into consideration the correspondence between the parties relating to the dispute and their respective efforts to arrive at a fair settlement.

Extracts from this clause have been quoted because they fairly indicate the substance of most arbitration clauses. An Arbitrator is always an experienced member of the trade and a panel of names is maintained by the Timber Trade Federation of the United Kingdom and other countries. Having chosen (an) Arbitrator(s) the parties await the reference or submission note which is prepared by the agent. This note follows a standard form and is a statement of the facts giving rise to the dispute. It details the buyer, shipper, reason for dispute and names the Arbitrator. The document is signed by both parties — the agent on behalf of his principal — and this together with all relevant correspondence is given to the Arbitrator(s). Under the Arbitration Act, the award must be made in writing within three months after commencement of arbitration, but this can be extended by mutual agreement if circumstances require further time.

4. AWARDS AND COSTS

When the award has been decided by the Arbitrator(s) and/or Umpire, it is recorded in writing in a standard form and the parties are informed that it is ready and will be delivered on payment of costs amounting to so much. It is customary for the buyer to "take up" the award by paying their costs. As mentioned earlier, these costs are entirely left to the discretion of the Arbitrator(s) and can be equally divided between the two parties or for the account of one party only. Depending on circumstances, it is not unknown for the buyer to pay costs exceeding the amount awarded to him—thus incurring a loss. Alternatively, the seller may have to pay all the costs in addition to the award.

The Arbitration Act, 1950, states: "An award on an Arbitration agreement may, by leave of the High Court or a judge thereof, be enforced in the same manner as a judgement or order to the same effect." This enforcement has the same status as a judgement, and by taking out an appropriate form of writ the applicant has the same rights but is limited by the jurisdiction of the Court, which does not extend to foreign countries.

Strictly speaking, there is no right of appeal against an award. Even if one party considers that a wrong decision has been made he must accept the award. Since the Courts have a general jurisdiction over Arbitrators and Umpires, this can be exercised within certain limits. For example, if one of the reasons for making the award is wrong in law or if there has been personal or legal misconduct, then the Courts can set the award aside. Personal misconduct can be any circumstance which mars the impartiality of the Arbitrator or Umpire such as concealment of facts or deception, fraud and bribery. It should be noted that such circumstances may be the fault of the parties concerned and not of the Arbitrators.

6

MENSURATION

1. GENERAL

Even in the seventeenth century timber was already being imported by the United Kingdom and there are recorded details of purchases following the Great Fire of 1666. These imports were both hardwoods and softwoods, although demands were also being satisfied from home-grown sources. In the eighteenth century the importation of both was firmly established and by the end of the century the softwood supplies were predominantly from the Baltic. It is an interesting fact that the form of measurement of Great Britain was adopted by all the exporting countries and has remained to this day. Undoubtedly, this was because Great Britain was the main customer of these exporting countries, which standardized their production accordingly. This has meant that exports of wood goods to countries other than the United Kingdom—those which normally use the metric system of measurement—have also been in imperial measure.

At the time of preparing this book—late 1968—plans have been made by the United Kingdom to completely adopt the metric system. This is being done on a long-term planned scale and the construction industry is due to change at the beginning of 1970. Since this industry is the major consumer of wood goods—particularly softwoods—it follows that the timber trade must change at the same time. By international agreement this has been planned, and exporting countries will be producing sawn goods in metric sizes in the second half of 1969 in order to export for 1970 consumption.

This quite revolutionary change to metrication after more than 200 years will inevitably bring with it many problems. There will be a transitional stage during which stocks will consist of two different units of measure. It is obvious that such an old-established trade will encounter many difficulties in accepting the new standards but it is equally obvious that it had to come if only in the interests of standardization throughout the world—lack of which has resulted in very considerable inefficiency and uneconomic situations over so many years.

In this chapter, it is proposed to refer to both imperial and metric measures in order to cover the transitional period. The older member of the timber trade will no doubt find it necessary to adopt "conversionary thinking" for some time, whereas the newer member will find the changeover less arduous.

2. IMPERIAL MEASURE

In general terms the lineal foot or running foot and the cubic foot are the present measurement units for all timber. The cubic foot is used by the hardwood trade and this is extended to a "standard" for the softwood trade—consisting of 165 cu. ft. The exception to this is the North American practice whereby a compromise is employed. The goods are measured in feet and inches but the volume is expressed in board feet—this being a square foot of timber 1 in. thick, i.e. one-twelfth of a cubic foot.

The origin of the standard as a unit of measure is associated with Russian exports. Known as the Petrograd Standard Hundred, abbreviated P.S.H., it consisted of 120 pieces (a "long hundred") measuring $1\frac{1}{2}$ in. by 11 in. and 12 ft long. The standard quantity of any specification is arrived at by obtaining the total lineal or running feet (addition of the various lengths), which is then multiplied by the width and thickness in inches and divided by 144. This gives total cubic feet which, divided by 165, gives total standards. Whilst the United Kingdom imports and resells softwood by the standard, there are a number of other units which are used for internal sales, the most important of which is possibly the square. This is a flooring measurement and consists of 100 sq. ft irrespective of thickness. The number of squares to a standard depends upon the thickness of the boards and is calculated on nominal measure. Since there are 1980 sq. ft of 1 in. material in a standard, 19·80 nominal squares (unplaned) equal 1 standard. It should be noted that when applied to planed boards it is sometimes qualified by the word "laid". A laid square is therefore the actual surface covered when the boards are laid, therefore the finished or "actual" width of the planed face of the board must be taken into consideration to allow for the difference between "nominal" and "laid" measure. It is customary for a planed board with a plain edge (p.e.) to finish $\frac{1}{4}$ in. less in width than the nominal or sawn width and for a planed, tongued and grooved (p.t.g.) board to finish $\frac{1}{2}$ in. less in width—the $\frac{1}{4}$ in. tongue being concealed in the groove.

The practice of using the standard and l.s.d. as our basis of measurement and payment has long involved the trade in tedious calculations. Our decision to change to the metric system is bound to create initial difficulties but should be welcomed by those with whom we have been out of step for so long.

3. METRIC MEASURE

The change from imperial to metric measure could not be made by merely converting the present common sizes from inches and feet to millimetres and metres. This can be illustrated by the fact that 1 in. equals 25·4 mm and since it is obviously desirable to use easily remembered sizes, then 25 mm has been adopted, which equals 0·98 in. This rule has been adopted throughout and, even though our commonly recognizable

71

sizes will look very much the same, there will be variations and these will be of particular significance during the transitional period of changeover. Quite obviously there will be a period of unknown length during which stocks in both exporters' and importers' yards—particularly the latter—will consist of both measurement units.

Because of the above-mentioned variations in conversion, there may be loss in volume when imperial is adjusted to metric and in many of the more popular sizes this will amount to as much as 4 per cent. Therefore it follows that particular attention must be given to reselling and associated price adjusting.

There has never been any accepted standardization in Europe for metric measure and this has resulted in a steady proliferation of sizes in commercial use over the years. The decision for a complete change from imperial to metric upon a national basis has provided the opportunity to adopt not only a rational range of sizes—particularly for softwood—but a greatly simplified one, with all the associated benefits that this can bring in terms of design, production and stockholding.

The sizes are related to, though not identical with, those which have been in common use for many years in imperial measure. The relationship is near enough in most instances to avoid the need for redesign. It must be recognized that there will be specialized end uses, and during the transitional period, there may be requirements for sizes not included amongst the simplified schedule of basic sizes. The supply of these will need to be negotiated between the parties concerned.

In so far as the United Kingdom is concerned, a new British Standard Specification has been issued. B.S. 4471—Dimensions for Softwood—Metric Units. The basic sawn sizes for softwood are shown in Tables VIII and IX. These sizes are to be measured as at 20 per cent moisture content. For any higher moisture content up to 30 per cent, the size will be greater by 1 per cent for every 5 per cent of moisture content in excess of 20 per cent and similarly adjusted for contents below 20 per cent.

TABLE VIII. BASIC SOFTWOOD SIZES—METRIC AND IMPERIAL MEASURES

Thickness		Width											
mm	in.	75	100	115	125	138	150	175	200	225	250	275	300
		3	4	4½	5	5½	6	7	8	9	10	11	12
16	5/8	X	X		X		X						
19	3/4	X	X		X		X						
22	7/8	X	X		X		X						
25	1	X	X		X		X	X	X	X	X		X
32	1¼	X	X		X		X	X	X	X	X		X
36		X	X		X		X						
38	1½	X	X		X		X	X	X	X			
40*		X	X		X		X	X	X	X			
44	1¾	X	X		X		X	X	X	X	X		X
50	2	X	X		X		X	X	X	X	X		X
63	2½		X		X		X	X	X	X			
75	3		X		X		X	X	X	X	X		X
100	4		X				X		X		X		X
150	6						X		X				X
200	8								X				
250	10										X		
300	12												X

* For 40 mm thickness, designers and users should check availability. The smaller sizes contained within the dotted lines are normally but not exclusively of European origin. The larger sizes outside the dotted lines are normally but not exclusively of North and South American origin.

The table has been extended to include commonly used widths not mentioned in the British Standard Specification (4½, 5½ and 11 in.).

TABLE IX. BASIC SOFTWOOD SIZES—METRIC AND IMPERIAL MEASURES

Lengths (0·30 m increments)

Metric (m)	Imperial (ft)
1·8	6
2·1	7
2·4	8
2·7	9
3·0	10
3·3	11
3·6	12
3·9	13
4·2	14
4·5	15
4·8	16
5·1	17
5·4	18
5·7	19
6·0	20
6·3	21

B.S. 4471—On lengths, no minus deviation is permissible, but over-length is unlimited.

Permissible tolerances in respect of dimension are given as follows:

Minus deviation in cross-section is permissible as not exceeding 10 per cent of the pieces in any parcel of sawn softwood.

Thickness and width—not exceeding 100 mm minus 1 mm plus 3 mm.

Thickness and width—over 100 mm minus 2 mm plus 6 mm.

Length—no minus deviation but over-length is unlimited.

Tables Xa, Xb and Xc show the new basic sizes together with the nominal and actual equivalents. These are given because they may be helpful in calculating losses or gains arising from conversion. Tables XII, XIII and XIV give the

TABLE Xa. BASIC SOFTWOOD SIZES—THICKNESS

Nominal		Actual	
Metric (mm)	Imperial (in.)	Metric (mm)	Imperial (in.)
16	0·625	15·875	0·630
19	0·750	19·050	0·748
22	0·875	22·225	0·866
25	1·000	25·400	0·984
32	1·250	31·750	1·261
36	1·438	36·525	1·418
38	1·500	38·100	1·500
40	1·625	41·275	1·576
44	1·750	44·450	1·773
50	2·000	50·800	1·970
63	2·500	63·500	2·482
75	3·000	76·200	2·955
100	4·000	101·600	3·940
150	6·000	152·400	5·910
200	8·000	203·200	7·800
250	10·000	254·000	9·850
300	12·000	304·800	11·820

1 mm : 0·0394 in.   1 in. : 25·4001 mm.

general conversion factors for imperial and metric units used by the Trade and Table XIV also gives the running metres in one cubic metre.

In processed softwood—finishing by planing—the permitted reductions from basic sizes are shown in Table XI. The processing to accurate finished sizes involves reduction of size varying with the size and the process. Exact figures cannot be established for each process since the original sawn size is not precise. After processing, a tolerance of plus or minus 0·5 mm is permitted.

TABLE Xb. BASIC SOFTWOOD SIZES—WIDTH

Nominal		Actual	
Metric (mm)	Imperial (in.)	Metric (mm)	Imperial (in.)
75	3·000	76·200	2·955
100	4·000	101·600	3·940
115	4·500	114·300	4·531
125	5·000	127·000	4·925
138	5·500	139·700	5·437
150	6·000	152·400	5·910
175	7·000	177·800	6·895
200	8·000	203·200	7·880
225	9·000	228·600	8·865
250	10·000	254·000	9·850
275	11·000	279·400	10·835
300	12·000	304·800	11·820

1 mm : 0·0394 in.   1 in. : 25·4001 mm.

TABLE Xc. BASIC SOFTWOOD SIZES—LENGTH

Nominal		Actual	
Metric (m)	Imperial (ft)	Metric (m)	Imperial (ft)
1·80	6	1·830	5·906
2·10	7	2·135	6·890
2·40	8	2·440	7·874
2·70	9	2·745	8·859
3·00	10	3·050	9·843
3·30	11	3·355	10·827
3·60	12	3·660	11·812
3·90	13	3·965	12·796
4·20	14	4·270	13·780
4·50	15	4·575	14·765
4·80	16	4·880	15·749
5·10	17	5·185	16·733
5·40	18	5·490	17·717
5·70	19	5·795	18·702
6·00	20	6·100	19·686
6·30	21	6·405	20·670

1 m : 3·281 ft.   1 ft : 0·3048 m.

TABLE XI. REDUCTIONS FROM BASIC SIZE TO FINISHED SIZE BY PROCESSING OF TWO OPPOSED FACES

Purpose	Sawn sizes—width or thickness in mm				
	15 to and including 22	Over 22 to and including 35	Over 35 to and including 100	Over 100 to and including 150	Over 150
Constructional timber surfaced	3	3	3	5	6
Floorings*	3	4	4	6	6
Matchings and interlocking boards*	4	4	4	6	6
Planed all round	4	4	4	6	6
Trim	5	5	7	7	9
Joinery and cabinet work	7	7	9	11	13

* The reduction of width is overall the extreme size and is exclusive of any reduction of the face by machining of a tongue or lap joint.

TABLE XII. LINEAR MEASURES—CONVERSIONS

Metres to Feet

(1 m = 3·280840 ft)

M →	0 ft	1 ft	2 ft	3 ft	4 ft	5 ft	6 ft	7 ft	8 ft	9 ft	M ←
0	—	3·281	6·562	9·843	13·123	16·404	19·685	22·966	26·247	29·528	0
10	32·808	36·089	39·370	42·651	45·932	49·213	52·493	55·774	59·055	62·336	10
20	65·617	68·898	72·178	75·459	78·740	82·021	85·302	88·583	91·863	95·144	20
30	98·425	101·706	104·987	108·268	111·549	114·829	118·110	121·391	124·672	127·953	30
40	131·234	134·514	137·795	141·076	144·357	147·638	150·619	154·199	157·480	160·761	40
50	164·042	167·323	170·604	173·885	177·165	180·446	183·727	187·008	190·289	193·570	50
60	196·850	200·131	203·412	206·693	209·974	213·255	216·535	219·816	223·097	226·378	60
70	229·659	232·940	236·220	239·501	242·782	246·063	249·344	252·625	255·906	259·186	70
80	262·467	265·748	269·029	272·310	275·591	278·871	282·152	285·433	288·714	291·995	80
90	295·276	298·556	301·837	305·118	308·399	311·680	314·961	318·241	321·522	324·803	90
100	328·084										100

Feet to Metres
(1 ft = 0.3048 m)

Feet	0	1	2	3	4	5	6	7	8	9	Feet
	m	m	m	m	m	m	m	m	m	m	
0	—	0·305	0·610	0·914	1·219	1·524	1·829	2·134	2·438	2·743	0
10	3·048	3·353	3·658	3·962	4·267	4·572	4·877	5·182	5·486	5·791	10
20	6·096	6·401	6·706	7·010	7·315	7·620	7·925	8·230	8·534	8·839	20
30	9·144	9·449	9·754	10·058	10·363	10·668	10·973	11·278	11·582	11·887	30
40	12·192	12·497	12·802	13·106	13·411	13·716	14·021	14·326	14·630	14·935	40
50	15·240	15·545	15·850	16·154	16·459	16·764	17·069	17·374	17·678	17·983	50
60	18·288	18·593	18·898	19·202	19·507	19·812	20·117	20·422	20·726	21·031	60
70	21·336	21·641	21·946	22·250	22·555	22·860	23·165	23·470	23·774	24·079	70
80	24·384	24·689	24·994	25·298	25·603	25·908	26·213	26·518	26·822	27·127	80
90	27·432	27·737	28·042	28·346	28·651	28·956	29·261	29·566	29·870	30·175	90
100	30·480										100

(Courtesy of *The Economist*.)

79

TABLE XIII. Cubic Measures—Conversions

Cubic Metres to Cubic Feet

(1 m³ = 35·31467 cu. ft)

m³ →	→ 0	1	2	3	4	5	6	7	8	9 ←	→ m³
	cu. ft	cu. ft	cu. ft	cu. ft	cu. ft	cu. ft	cu. ft	cu. ft	cu. ft	cu. ft	
0	—	35·31	70·63	105·94	141·26	176·57	211·89	247·20	282·52	317·83	0
10	353·15	388·46	423·78	459·09	494·41	529·72	565·03	600·35	635·66	670·98	10
20	706·29	741·61	776·92	812·24	847·55	882·87	918·18	953·50	988·81	1,024·13	20
30	1,059·44	1,094·75	1,130·07	1,165·38	1,200·70	1,236·01	1,271·33	1,306·64	1,341·96	1,377·27	30
40	1,412·59	1,447·90	1,483·22	1,518·53	1,553·85	1,589·16	1,624·47	1,659·79	1,695·10	1,730·42	40
50	1,765·73	1,801·05	1,836·36	1,871·68	1,906·99	1,942·31	1,977·62	2,012·94	2,048·25	2,083·57	50
60	2,118·88	2,154·19	2,189·51	2,224·82	2,260·14	2,295·45	2,330·77	2,366·08	2,401·40	2,436·71	60
70	2,472·03	2,507·34	2,542·66	2,577·97	2,613·29	2,648·60	2,683·91	2,719·23	2,754·54	2,789·86	70
80	2,825·17	2,860·49	2,895·80	2,931·12	2,966·43	3,001·75	3,037·06	3,072·38	3,107·69	3,143·01	80
90	3,178·32	3,213·63	3,248·95	3,284·26	3,319·58	3,354·89	3,390·21	3,425·52	3,460·84	3,496·15	90
100	3,531·47										100

Cubic Feet to Cubic Metres
(1 cu. ft = 0.028317 m³)

Cu. ft →	→ 0	10	20	30	40	50	60	70	80	90 ←	Cu. ft →
	m³	m³	m³	m³	m³	m³	m³	m³	m³	m³	
0	—	0·283	0·566	0·850	1·133	1·416	1·699	1·982	2·265	2·549	0
100	2·832	3·115	3·398	3·681	3·964	4·248	4·531	4·814	5·097	5·380	100
200	5·663	5·947	6·230	6·513	6·796	7·079	7·362	7·646	7·929	8·212	200
300	8·495	8·778	9·061	9·345	9·628	9·911	10·194	10·477	10·760	11·044	300
400	11·327	11·610	11·893	12·176	12·459	12·743	13·026	13·309	13·592	13·875	400
500	14·159	14·442	14·725	15·008	15·291	15·574	15·858	16·141	16·424	16·707	500
600	16·990	17·273	17·557	17·840	18·123	18·406	18·689	18·972	19·256	19·539	600
700	19·822	20·105	20·388	20·671	20·955	21·238	21·521	21·804	22·087	22·370	700
800	22·654	22·937	23·220	23·503	23·786	24·069	24·353	24·636	24·919	25·202	800
900	25·485	25·768	26·052	26·335	26·618	26·901	27·184	27·467	27·751	28·034	900
1000	28·317										1000

(Courtesy of *The Economist.*)

81

TABLE XIV. RUNNING METRES IN 1 m³

Thickness (mm)	Width (mm)									
	75	100	125	150	175	200	225	250	275	300
16	833·33	625·00	500·00	416·67	357·14	312·50	277·77	250·00	227·27	208·33
19	701·75	526·32	421·05	350·87	300·75	263·16	233·92	210·52	191·39	175·44
22	606·06	454·54	363·64	303·03	259·74	227·27	202·02	181·82	165·29	151·51
25	533·33	400·00	320·00	266·67	228·57	200·00	177·78	160·00	145·45	133·33
32	416·66	312·50	250·00	208·33	178·57	156·25	138·89	125·00	113·64	104·17
38	350·88	263·16	210·53	175·44	150·38	131·58	116·96	105·27	95·69	87·72
44	303·03	227·27	181·81	151·52	129·87	113·64	101·01	90·91	82·64	75·76
50	266·66	200·00	160·00	133·33	114·29	100·00	88·89	80·00	72·73	66·67
63	211·64	158·73	126·98	105·82	90·70	79·36	70·55	63·49	57·72	52·91
75	177·78	133·33	106·67	88·89	76·19	66·67	59·26	53·33	48·48	44·44
100	133·33	100·00	80·00	66·67	57·14	50·00	44·44	40·00	36·36	33·33

1 mm = 0·0394 in. 1 in. = 25·4001 mm.
1 m = 39·3701 in. 1 ft = 304·8010 mm.
1 m = 3·2808 ft. 1 ft = 0·3048 m.
1 m² = 10·7639 sq. ft. 1 sq. ft = 0·0929 m².
1 m³ = 35·3147 cu. ft. 1 cu. ft = 0·0283 m³.
1 m³ = 0·214 standard. 1 standard = 4·6720 m³.

ALLIED TRADE PRODUCTS
AND DEVELOPMENTS

1. GENERAL

The preceding chapters have dealt, in the main, with timber—
both softwoods and hardwoods. In this chapter, allied products
such as plywood, fibre building boards and particle boards
will be discussed together with some developments which are
creating quite revolutionary changes in the trade as a whole—
these being packaging, finger jointing and kiln drying.

2. WOOD-BASED PANEL PRODUCTS

This term broadly covers four materials—plywood (including blockboard), particle board, compressed fibreboard (hardboard), non-compressed fibreboard (insulation board). There
is considerable overlapping of the end-uses of these materials
and, for this reason, it is often desirable to regard them as a
group, i.e. for statistics showing consumption, etc. Choice
of one product rather than another may be guided by habit,
price or sales promotion and advertising. Because of this
constantly changing pattern of use, grouping is inevitable since
there is no definite limit to end-use of any one individual
product.

The 1950's showed remarkable growth in the consumption
of panel products. In Europe alone, this rose by over 200 per
cent between 1950 and 1960 and it is estimated that between

83

then and 1975 the figure will treble (see Table XV). This is a greater rise than can be expected for sawn goods and the main reason for this is as follows. Panel products have the capacity to replace natural wood goods in so many of the fields which have always been traditional. A relatively small amount of time and labour is needed to convert wood panels into a finished product. If a comparison is made between panels and their main rival, sawn goods, it is found that they are less wasteful, less demanding in skilled labour, quicker to handle and fix in place, and need fewer finishing processes. There is little doubt that product development will widen the field of use and that there will be a continual encroachment on sawn goods in many industries. In a decade when most materials rose in price, panel products actually became cheaper.

TABLE XV. EUROPEAN CONSUMPTION OF PANEL PRODUCTS
(Thousand cubic metres)

Region	1950	1960	1975 estimated
Northern Europe	462	926	1720
E.E.C.	1028	3434	9080
British Isles	440	1281	3840
Central Europe	108	489	2110
Southern Europe	64	192	700
Eastern Europe	248	1203	5150
Total	2350	7525	22,600

Table XVI shows the caput consumption of panel products in Europe. It will be seen that there is a very marked difference between the northern and southern regions. Countries which have, in the past, been considerable users of sawn goods offer the greatest field for substitution, whereas regions such as Southern Europe—where the use of wood in any form is low— have a correspondingly low consumption of panel products.

TABLE XVI. EUROPEAN CONSUMPTION OF PANEL PRODUCTS
(Cubic metres per thousand capita)

Region	1950	1960	1975 estimated
Northern Europe	27·7	48·8	81·2
E.E.C.	10·0	24·1	57·1
British Isles	9·8	25·4	69·9
Central Europe	5·5	17·8	63·7
Southern Europe	1·1	2·7	7·8
Eastern Europe	3·7	13·8	50·8
Total	7·6	19·4	49·7

In building and construction, in furniture and packaging, the panel products have replaced timber to a very considerable degree. Sheathing for timber frame houses, roofing, flooring and concrete shuttering or formwork is now common practice. The majority of our doors are faced with plywood or hardboard and boxboards are being replaced in the packaging industry. The development of completely water-resistant adhesives has done much to encourage this utilization since the resultant durability has opened up so many new fields where exposure to adverse conditions is common, of which the boat-building industry is an example.

3. PLYWOOD

In its commonest form, it is made by combining three sheets of veneer, the grain of the central veneer being at right angles to that of the outer sheets. The sheets are then glued together under pressure. Two, three, five, seven or more veneers may be used and there is a wide variety of plywoods available, differing in thickness, in strength, in durability and in appearance.

It is a panel material having remarkably high strength properties for its weight. It is dimensionally stable and the

85

development of synthetic resin adhesives has made it possible to produce completely waterproof grades. Where strength and lightness are of primary importance, its position is unchallenged.

Associated with plywood is blockboard and, whilst its construction is different, the two are combined for statistical purposes and they are usually manufactured by the same mills. Blockboard differs from plywood in that its centre or core is composed of strips of softwood up to 1 in. thick which are glued together. This core is faced with veneer or plywood to form a thick and extremely rigid sheet. It is this material which receives the greatest competition from particle board—which is described later.

Commercial plywood is normally obtainable in thicknesses ranging from $\frac{1}{8}$ in. to 1 in. (3 mm–25 mm) but this range can be extended for special end-uses. The bonding media used in the manufacture of plywood are of the utmost importance since their properties determine the characteristics and end-usage of the final product. For example, plywood used for internal application such as furniture needs to be well bonded but does not need to be highly resistant to moisture or water. Conversely, plywood used externally and exposed to adverse weather conditions requires an adhesive of high durability. Therefore, it is essential for the user to understand the properties of the many types of adhesives since it is mainly this aspect which determines the suitability of the material for a particular end-use. In very broad terms the following types of adhesive are used in order of durability:

Interior Use. Animal glues, blood albumen, casein and soya.

Moisture Resistant (M.R.). Urea–formaldehyde (U.F.).

Boil Resistant (B.R.). Urea–melamine formaldehyde (U.M.F.).

Weather and Boil Proof (W.B.P.). Phenol–formaldehyde (P.F.) and resorcinol–formaldehyde (R.F.)

All British-made and most imported plywood is manufactured with synthetic resin adhesives in one form or another. It should be noted that synthetic resins can be extended to

TABLE XVII. PRODUCTION OF PLYWOOD

(Thousand cubic metres)

Region	1950	1960	1968
Europe	1,315	2,680	3,565
U.S.S.R.	658	1,353	1,832
N. America	3,635	8,910	15,009
C. America	10	40	133
S. America	210	205	358
Africa	50	130	216
Asia	215	1,940	6,812
Pacific Area	80	155	136
World Total	6,173	15,413	28,061

87

such a degree that they lose their moisture-resistant properties and may only be suitable for interior work. Extenders or adulterants are normally in powder form and are included to increase the spread of the adhesive, so reducing cost. The term "resin-bonded plywood" is in itself valueless since adulteration can cause it to fall below the requirements of M.R.-type adhesives.

The following countries are major manufacturers and exporters of plywood: Austria, Bulgaria, Canada, Congo Republic, Czechoslovakia, Denmark, Finland, France, Gabon (West Africa), Germany, Israel, Japan, Nigeria, Romania, Surinam (Dutch Guiana), Sweden, the United Kingdom, the United States, the Soviet Union, Yugoslavia. All these countries have their own grading rules which clearly define the qualities and properties of the product (Table XVII).

4. FIBRE BUILDING BOARDS

The basic characteristic of all fibreboards is that they are made from wood fibres, i.e. woodpulp. Since wood fibres are self-adhesive, most fibreboards are made without any additional bonding material. There is a wide variety of products all of which originate from wood waste or roundwood which has been subjected to a pulping process to form a mat which is then pressed to form a board. When this pressure is light, the product is known as insulation board (non-compressed), and when high pressure and heat are applied the product is known as hardboard (compressed). These are broad distinctions, and a wide variety of strength and densities can be obtained within the range of the two materials, together with final treatments such as impregnation and tempering.

Approximately 60 per cent of the total capacity of the fibreboard industry is located in Northern Europe. This is related to the fact that some 80 per cent of the northern mills are integrated and are thus able to obtain a substantial part of their raw material requirements in the form of residues or waste from the parent sawmill. Apart from this source of raw material, roundwood is used, the small logs being unsuitable as saw logs. This means that the board mill must compete with the pulp, paper and the particle board industry for its raw material. A conservative estimate is that Europe as a whole uses 65 per cent wood residue and 35 per cent round-wood for the fibreboard industry.

Insulation Board

This type of board is soft and porous and has relatively little strength. It is predominantly used for its excellent insulating and acoustic properties and it is an ideal inexpensive building board. In latter years, considerable research has resulted in fire-retarding treatments and this is undoubtedly increasing the popularity of the material, some of which comes within the requirements of present-day building regulations—these giving emphasis to fire, particularly in multistorey buildings.

Insulation board is obtainable in a variety of thicknesses from $\frac{1}{2}$ in. up to 1 in. and choice of sheet size is very wide. Small tiles to large sheets—the latter only limited by handling difficulty—can be produced with a variety of surface finishes and patterns.

Hardboard

This material can be regarded as the most versatile of all the fibreboards. It would be quite impossible to list all the many uses employed all over the world. There is infinite variety of densities, the most commonly styled grades being "standard", with a softer grade called "medium", and a harder grade called "tempered". Thickness ranges from $\frac{3}{32}$ in. to a maximum of $\frac{1}{2}$ in.

Hardboard is heavier than plywood but, where lightness is not a prerequisite, its cheapness and the ease with which it takes surface finishes give it many advantages. It is used in

89

furniture manufacture, for flush doors and similar panels, packaging, display boards, vehicle construction and as a flooring material—especially as an underlay. The tempered grade is also used for concrete shuttering or form-work and for exterior cladding. It is probably the most widely used of all boards in the very significant "do-it-yourself" market.

Apart from density classification, it is available in a very wide variety of surface finishes. It can be pre-enamelled or lacquered in any colour in plain or tiled pattern, embossed or moulded to simulate leather, reeded or fluted patterns. In recent years, the demand for perforated hardboard has increased considerably—it being widely used for display purposes.

5. PARTICLE BOARDS

This material can be said to be the latest addition to the family of panel products. The first commercial plant started in Germany in 1941 and by 1949 the extrusion process was developed. It really belongs to the post-war years and its development and utilization have been both rapid and widespread throughout the world, where it is also known as chipboard and flakeboard. It is made by combining wood chips, flakes or shavings with a synthetic resin adhesive and then bonding the mass together under heat and pressure. This produces a thick, stable and rigid sheet. It can be easily worked, can be given a variety of finishes and, above all, it has outstanding dimensional stability.

Particle board is obtainable in a variety of thicknesses ranging from $\frac{3}{8}$ in. up to 1 in. The continuous process for flat-pressed boards has meant that there is no theoretical limit to the length of the board but, as with other panel products, the problems of handling are usually the limiting factor. Because it is a homogeneous material, it is less affected by humidity changes and the dimensional movement is small and is equal in all directions—unlike other wood-based materials. Such uniformity of movement eliminates the distortion normally

TABLE XVIII. PRODUCTION OF PARTICLE BOARD
(Thousand metric tons)

Region	1955	1960	1968
Europe	290	1200	5108
U.S.S.R.	—	98	979
N. America	120	355	1792
C. America	—	—	68
S. America	5	22	133
Africa	30	13	39
Asia	5	80	318
Pacific Area	—	6	140
World Total	450	1774	8577

PRODUCTION OF FIBRE BOARD
(Thousand metric tons)

Region	1955	1960	1968
Europe	1190	1740	2701
U.S.S.R.	54	214	548
N. America	1645	1815	2716
C. America	10	20	41
S. America	25	75	140
Africa	70	135	80
Asia	45	105	623
Pacific Area	110	145	201
World Total	3149	4249	7050

associated with natural timber. Additionally, this uniformity and freedom from grain direction enable the board to be sawn and machined in any direction with the minimum of waste.

Control in manufacture provides a wide range of densities and grades suitable for an ever-increasing variety of uses. The process of manufacture can be changed to meet almost any requirement and an example of this is the multilayer board. This may be made from wood particles of varying sizes, graded outwards from the centre of the board, with the finest particles towards the outer surfaces. This provides an exceptionally smooth surface which takes a decorative finish— often embodied in the manufacturing process.

In the building industry, particle board is established as a flooring material either on joists or on a solid sub-floor and as dry wall partitioning. In the latter use, factory-finished boards with ready-to-paint surfaces and facings of aluminium and asbestos are available. Fire-retardant properties can also be introduced to meet present-day regulations.

The furniture industry was the original consumer of particle boards and was quick to appreciate its qualities. Faced with decorative wood veneers or plastic facings it provided superior panels of strength, stiffness and, above all, dimensional stability. Its adaptability for all types of machining such as dovetails, tongues and grooves, screw holding and gluing properties have accounted for the almost phenomenal increase in consumption. In Europe alone, this rose from 1,488,000 tons in 1961 to 3,342,000 in 1965. It is estimated that in 1975, this figure will rise to 6,300,000 tons. See Table XVIII.

6. PACKAGED TIMBER

The growth of this system of delivering and handling timber has been one of the most significant developments of recent years in the timber trade. Some ten years ago, timber was being shipped by exporting countries as loose pieces—as it had been since the beginning of the trade. The adoption of this system was the logical outcome of the need to effect economies in handling or, in other words, to reduce labour costs and the time factor.

Packaging of timber consists of standardized bundles of pieces, all of one dimension and usually of one length. The package is bonded with steel or nylon bands—the number depending on the length of the package. The size of the package has been the subject of lengthy discussion on an international basis. Broadly speaking, opinion is divided between the Canadian preferred system and that of the Baltic countries—this will be discussed later.

Because the handling methods for timber have become uneconomic in the light of present-day high labour costs, it follows that it was imperative to devise ways and means of streamlining the system. The costs of loading and discharging, stowing in ships' holds, transportation after discharge and storage in the receiver's yard can all be effectively reduced. The average general cargo ship gains a quicker turn-round at ports, thus reducing freight rates, and this is of great significance since most ships spend some 60 per cent of their time in ports where no income is earned for the owners. Packaged goods can be mechanically handled both during loading and discharging and the receiver can minimize his piling costs since his goods are already sorted to length. The ultimate object should be that there would arrive at the dockside a unit load which can be easily tallied and which could be successively rehandled up to the final point of delivery in the most economical manner. In theory, this appears to be the ideal solution to all problems, but in practice it has presented some importers with additional ones. This particularly applies to the smaller concerns which are not adequately equipped with handling facilities such as forklift trucks of sufficient capacity to cope with the packages.

At this point it is desirable to explain the basic difference between the approach to packaging by the Canadians and the Central and Northern European exporters. Exporters on the west coast of Canada—who represent a very large proportion of the world's available softwood—have used liberty ships until these became no longer serviceable. The bulk carrier was then introduced which was designed to carry bulk cargoes in easily handled form. Canada established the size of package best suited to these bulk carriers and, needless to say, the large unit proved to be the most economical to handle. Ships capable of carrying as much as 7500 standards, loaded and discharged by their own gantry cranes, have become commonplace. This has meant that the number of receiving ports has become fewer since they could not provide facilities for such large ships. This led to the development of terminals especially designed to cope with the handling of packaged units and deep draught vessels. In so far as the United Kingdom is concerned it has meant that the centres of intake have been reduced down to about 25 per cent of the number of traditional timber receiving centres. In order to service those receivers who are located at some distance from such terminals and who may not wish to receive bulk cargoes (i.e. fewer large quantities rather than frequent small quantities) a new pattern of trading has emerged. Large storage areas have been created from which smaller consignments can be delivered by road and rail as required. It is important to remember that not only is the Canadian package of larger dimension: it is also heavier than those from other sources because its contents have not been seasoned. The ultimate receiver is therefore faced with the handling of a package which could weigh as much as 4 tons—very often beyond the capabilities of the smaller concern.

The European trend is quite different in that smaller ships and packages are favoured. Of necessity, this arises from the fact that the Baltic ports are not capable of coping with bulk carriers and this in turn means that the traditional smaller capacity ships can continue to ply between the Baltic and the very many receiving ports which cannot handle bulk carriers. Thus it would seem that the best of two worlds is satisfied by geographical considerations. However, the Baltic countries are fully aware of the necessity to provide packaged goods and the introduction of length sorting and packaging machines has become commonplace. Small but specially designed ships having shallow draught and with square holds and one large hatch will ensure speedy turn-round with resultant reduction in freight rates.

The difference between package sizes from Canada and Europe does not mean that there has been no agreement on an international basis. Overall maximum dimensions have been agreed, these being, in the main, related to the basic dimensions of timber-carrying transport, both road and rail.

The maximum recommended width of a package is approximately 44 in., which allows two side by side on a standard road vehicle of 7 ft 6 in. width. This will also allow the preferred European small package of 22 in. to occupy the same area but with four packages side by side—there being sufficient tolerance to allow for size discrepancies.

In conclusion, the all-important question of deterioration in the package form must be mentioned. The West Coast Canadian exporters enjoy the advantage of having species which do not discolour to the same extent as European softwoods, and they can therefore ship the timber in green condition and save the expense of seasoning. Knowing that this is the custom, receivers normally have to open packages on arrival and pile for seasoning before the timber is in suitable condition for end-use.

Conversely, the Baltic exporters in Russia, Finland and Sweden are required to package the goods in shipping dry condition. It follows that receivers tend to store the packages as received and it could be many months before they are opened and used. Despite the use of anti-stain treatments, it is still possible that close confinement in the package could lead to deterioration of the internal contents. For this reason, more than ever before, the Baltic exporters are concentrating on the need to supply adequately seasoned goods.

95

To illustrate the significance of packaging in the United Kingdom alone, the importation in 1963 was some 15,000 standards. By 1965 this rose to 105,000 standards, the next year to 300,000 and in 1967 some 450,000 standards were imported. If one estimates the total annual import of softwoods to be 1,800,000 standards this amounts to a total of possibly 306,000,000 pieces of timber. If this could be packaged in, say, four packages to the standard, then the total units to be handled would be reduced to 7,200,000. If these units are multiplied by the number of times they are handled between the exporter's mill and the ultimate end-user, then it becomes abundantly clear that economies can be effected.

7. FINGER-JOINTING

There are three main advantages to the end-jointing of timber. The limitation of available length is eliminated, the quality of low-grade timber can be upgraded by removal of defects such as knots, and short lengths which would normally be discarded as waste can be converted into usable material.

The conventional finger-joint is produced by cutting tapered fingers on the ends of the pieces to be joined, interlocking and gluing the two pieces together under pressure. By varying the length and pitch of these fingers and the width of their tips, joints of different characteristics can be produced. As a general rule, the longer the fingers and the narrower the tips, the stronger the joint. It has been proved that a joint with fingers of 2 in. long with a pitch of $\frac{7}{16}$ in. and a tip width of $\frac{1}{16}$ in. will be sufficiently strong for all structural applications. Joints with a finger of about 1 in. long have adequate strength for most joinery purposes. Figure 3 shows the general form of such joints.

Returning to the advantages to be gained by finger-jointing, these should be further clarified in the light of present-day developments in the trade as a whole. The removal of length limitation is particularly important in the field of structural laminated timber used for long-span beams, arches and roof-trusses. The use of long-length members brings the prospects for timber engineering techniques nearer to the competition of other structural materials. End-jointing techniques introduced at the exporter's sawmill can effect the production of high-grade material by cutting out the worst defects and so producing clear grades of associated higher value. The small

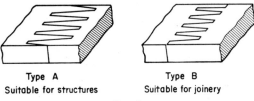

Type A
Suitable for structures

Type B
Suitable for joinery

FIG. 3

amount of cut-away waste can be better utilized by the exporter—especially if the mill is integrated. Additionally, length control resolves many of the problems arising with packaged timber in a limited range of pre-cut lengths.

From the end-use point of view, the advantage of reclaiming short offcuts is very considerable. In joinery works, the wide variety of components creates an inevitable source of such pieces which are normally wasted. It will be seen that both the exporter and the importer can benefit from this form of end-jointing—which has proved to be superior to any other form such as butt and scarf joints.

There are now available many types of fully automated machines for producing finger-joints. These can be fed with a continuous supply of pieces of timber of equal moisture content. The machine cuts the matching fingers, applies the glue to the joint, sets the glue by high-frequency heating and ejects a continuous piece which can be automatically crosscut to the desired length. Crosscutting facilities before the piece enters the machine can remove major defects such as knots, etc. The glue most commonly used is of the P.F. and R.F. types, so giving the highest degree of durability (see section 3, Plywood).

97

8. KILN DRYING

It is not intended to describe the various types of kilns, but rather to outline the advantages of seasoning timber in a shorter time and to a lower moisture content than can be done by air seasoning. The British Standards definition of seasoning is "a process involving the reduction of the moisture content in timber towards or to an amount suitable for the purpose for which it is to be used". In effect, this means bringing the timber to a condition in which it will remain reasonably stable and give satisfactory service after manufacture. Being hygroscopic, timber will always reach a state of equilibrium with surrounding atmospheric conditions, and unless this is fully appreciated, it is inevitable that unsatisfactory service will result—with accompanying disservice to the timber trade as a whole. In terms of service, there is no doubt that the most significant factor is seasoning, since the most serious faults which arise can usually be attributed to the moisture content of the timber. Seasoning is essential because it gives the following advantages to all species—whether softwood or hardwood:

(a) It becomes dimensionally stable—i.e. differential movement is reduced.

(b) It becomes lighter and stronger.

(c) It becomes resistant to discoloration and decay—all forms of which demand moisture above a certain level.

(d) It permits the better application of paints and other surface finishes.

(e) It permits the application of preservatives because better absorption is obtained.

(f) It permits a better surface finish when planing or machining.

Table XIX indicates the appropriate moisture content for different applications.

TABLE XIX. MOISTURE CONTENT OF TIMBER FOR VARIOUS PURPOSES

Moisture content (per cent)	Description
27	Appreciable shrinkage starts at about this point
26	
25	Suitable moisture content for pressure treatment — creosoting, fire resistance
24	
23	Carcassing timbers
22	} Range of moisture content attained in thoroughly air-seasoned timber
21	}
20	Dry rot safety line
19	
18	External joinery
17	External doors, agricultural implements and garden furniture
16	Motor vehicles, patterns, ships' decking and timber for general internal joinery
15	Bedroom furniture and wood in buildings slightly or occasionally heated
14	Internal doors and wood in buildings with regular intermittent heating
13	
12	Joinery and other wood in continuously heated buildings, e.g. furniture, block floors, etc.
11	
10	Woodwork in situations with high degree of central heat, e.g. hospitals, etc.
9	
8	Wood flooring over heating elements
7	
6	
5	
4	
3	
2	
1	
0	Oven-dried wood

Air drying sufficient (approx. 23–20)

Artificial heat necessary to secure sufficient drying (approx. 9–0)

(Courtesy of Timber Research and Development Association.)

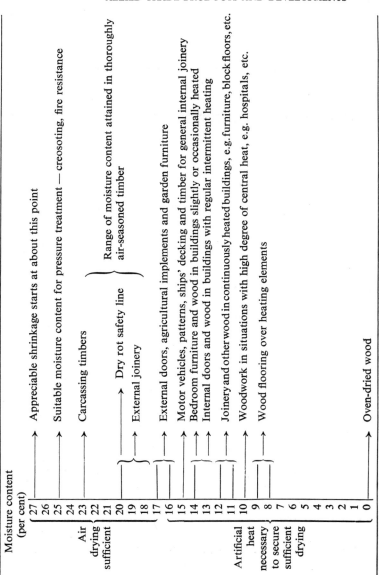

99

It will be seen that it is necessary to kiln dry timber if a moisture content of below about 17 per cent is required. This figure may vary slightly depending upon the climatic conditions of the country in question and upon the time of year. It is quite impossible to air-season timber in the open to a moisture content below the prevailing equilibrium conditions despite the fact that one hears it said, particularly of hardwoods, that it is "well seasoned after X years in stock". In the United Kingdom and throughout Europe, all timber for at least interior use should be used at a moisture content which can usually be attained only by artificial means—kiln drying.

It is unfortunately true to say that, not only in the past but even in present times, there has been insufficient attention paid to this very important question of seasoning. One is constantly seeing examples of incorrect or non-existent seasoning, particularly in joinery and interior woodwork, which has been installed at too high a moisture content and has shrunk and distorted when heating has dried out the building. All too common a sight is pre-manufactured joinery piled in the open in all weathers awaiting installation. Undoubtedly this has contributed to the loss of many traditional uses for wood to other processes, of which metal windows and composite floors are two examples.

Apart from the use of drying kilns by the utilization countries, there has been a definite trend towards their use in exporting countries.

In these instances the requirement is different since the timber is exported "shipping dry" and not necessarily ready for any particular end-use. As mentioned in a previous chapter, exporters are required to "properly season for shipment to the receiving country". This seldom means that a moisture content of less than 20 per cent is attained—which can normally be achieved by air seasoning. However, in recent years, two factors have introduced the desirability of kiln drying—particularly in the softwood-exporting countries of the Baltic. These are economics and packaging.

The long-established practice of air drying has involved very considerable labour employed to open pile in the open and to unpile at the end of the seasoning period. Since labour costs have continually risen in all countries, there has been need to reduce it whenever possible. Additionally, the exporter has had to endure the high cost of capital which has been unprofitably "frozen" by the goods being piled over periods of many months. The ability to produce seasoned goods to the required specification in a matter of days rather than months and to export promptly has obvious economic advantages.

The packaging of timber has necessitated greater need for adequate seasoning, as explained in section 6, Packaged Timber. In order to protect themselves against complaints of deterioration within the close confinement of timber packages, exporters appreciate the need to season to a slightly lower moisture content than has been usual hitherto. This does not necessarily mean that goods will be dried down to end-use moisture contents, but rather that the distribution of moisture within the piece will be more even and that the danger of internal moisture movement to the surface is eliminated. This can only be achieved by controlled drying, which is virtually impossible if one is dependent upon fluctuating and unpredictable climatic conditions.

ABBREVIATIONS COMMONLY USED IN THE TIMBER TRADE

a.d.	Air dried	f.o.t.	Free on truck
av. len.	Average length	F.O.W.	First open water
bd.	Board	f.p.a.	Free from particular average
bd. ft	Board foot		
bdl.	Bundle	f.s.p.	Fibre saturation point
B/E	Bill of entry		
B/L	Bill of lading	ft	Foot
b.m.	Board measure	hdwd.	Hardwood
B.R.	Boil resistant	h.g.	Home grown
b.s.	Band sawn	K.D.	Kiln dried
C. & B.	Clear and better	lgth.	Length
c.i.f.	Cost, insurance, freight	lin. ft	Lineal foot
		lin. m	Lineal metre
Clr.	Clear	M	Thousand
Com.	Common	m	Metre
C/P	Charter party	m³	Cubic metre
cu. ft	Cubic foot	M.b.m.	Thousand board measure
d.b.b.	Deals, battens, boards		
		m.c.	Moisture content
d.w.	Dead weight (shipping)	Merch.	Merchantable
		mm	Millimetre
E.G.	Edge grain	M.R.	Moisture resistant
e.m.	End matched	N.A.A.	Not always afloat
e.m.c.	Equilibrium moisture content	nom.	Nominal
		p.a.r.	Planed all round
f.a.q.	Fair average quality	PCP	Pentacholophenate
F.A.S.	Firsts and Seconds	pcs.	Pieces
f.a.s.	Free alongside	p.e.	Plain-edged
f.b.m.	Foot, board measure	P.F.	Phenol–formaldehyde
f.e.	Feather-edged		
f.o.b.	Free on board	p.s.e.	Planed and square-edged
f.o.l.	Free on lorry		
f.o.q.	Free on quay	P.S.H.	Petrograd Standard Hundred
f.o.r.	Free on rail		

p.t.g.	Planed, tongued and grooved	S.1.S.1.E.	Surfaced one side and one edge
q.g.	Quarter girth	S.2.S.1.E.	Surfaced two sides and one edge
Qtd.	Quartered		
R.F.	Resorcinol–formaldehyde (adhesive)	S.1.S.2.E.	Surfaced one side and two edges
r.l.	Random length	t. & g.	Tongued and grooved
r.s.	Rotary sawn	t.g.b.	Tongued, grooved and beaded
run. ft	Running foot		
run. m	Running metre	t.g.v.	Tongued, grooved and V-jointed
s.e.	Square-edged	t. & t.	Through and through (sawing)
sftwd.	Softwood		
S.g.	Specific gravity	u.b.	Under bark
Sh.D.	Shipping dry	u/e	Unedged
s.n.d.	Sap no defect	U.F.	Urea–formaldehyde (adhesive)
sq. ft	Square foot		
Std.	Standard	U.M.F.	Urea-melamine formaldehyde (adhesive)
sup. ft	Superficial foot		
S.1.E.	Surfaced one edge	U/S	Unsorted
S.2.E.	Surfaced two edges	W.B.P.	Weather and boil proof
S.1.S.	Surfaced one side (face)		
S.2.S.	Surfaced two sides	W/E	Waney-edged
		wt.	Weight

INDEX